Springer Theses

Recognizing Outstanding Ph.D. Research

Aims and Scope

The series "Springer Theses" brings together a selection of the very best Ph.D. theses from around the world and across the physical sciences. Nominated and endorsed by two recognized specialists, each published volume has been selected for its scientific excellence and the high impact of its contents for the pertinent field of research. For greater accessibility to non-specialists, the published versions include an extended introduction, as well as a foreword by the student's supervisor explaining the special relevance of the work for the field. As a whole, the series will provide a valuable resource both for newcomers to the research fields described, and for other scientists seeking detailed background information on special questions. Finally, it provides an accredited documentation of the valuable contributions made by today's younger generation of scientists.

Theses are accepted into the series by invited nomination only and must fulfill all of the following criteria

- They must be written in good English.
- The topic should fall within the confines of Chemistry, Physics, Earth Sciences, Engineering and related interdisciplinary fields such as Materials, Nanoscience, Chemical Engineering, Complex Systems and Biophysics.
- The work reported in the thesis must represent a significant scientific advance.
- If the thesis includes previously published material, permission to reproduce this must be gained from the respective copyright holder.
- They must have been examined and passed during the 12 months prior to nomination.
- Each thesis should include a foreword by the supervisor outlining the significance of its content.
- The theses should have a clearly defined structure including an introduction accessible to scientists not expert in that particular field.

More information about this series at http://www.springer.com/series/8790

Naoya Kanazawa

Charge and Heat Transport Phenomena in Electronic and Spin Structures in B20-type Compounds

Doctoral Thesis accepted by
the University of Tokyo, Tokyo, Japan

Springer

Author
Dr. Naoya Kanazawa
Department of Applied Physics
The University of Tokyo
Tokyo
Japan

Supervisor
Prof. Yoshinori Tokura
Department of Applied Physics
The University of Tokyo
Tokyo
Japan

ISSN 2190-5053 ISSN 2190-5061 (electronic)
Springer Theses
ISBN 978-4-431-56306-8 ISBN 978-4-431-55660-2 (eBook)
DOI 10.1007/978-4-431-55660-2

Springer Tokyo Heidelberg New York Dordrecht London

Softcover reprint of the hardcover 1st edition 2015

Printed on acid-free paper

Springer Japan KK is part of Springer Science+Business Media (www.springer.com)

Parts of this thesis have been published in the following journal articles:

1. Chapter 4: "Large topological Hall effect in a short-period helimagnet MnGe", N. Kanazawa, Y. Onose, T. Arima, D. Okuyama, K. Ohoyama, S. Wakimoto, K. Kakurai, S. Ishiwata, and Y. Tokura, Phys. Rev. Lett. **106**, 156603 (2011).
2. Chapter 4: "Possible skyrmion-lattice ground state in the B20 chiral-lattice magnet MnGe as seen via small-angle neutron scattering", N. Kanazawa, J.-H. Kim, D. S. Inosov, J. S. White, N. Egetenmeyer, J. L. Gavilano, S. Ishiwata, Y. Onose, T. Arima, B. Keimer, and Y. Tokura, Phys. Rev. B **86**, 134425 (2012).
3. Chapter 5: "Discretized topological Hall effect emerging from skyrmions in constricted geometry", N. Kanazawa, M. Kubota, A. Tsukazaki, Y. Kozuka, K. S. Takahashi, M. Kawasaki, M. Ichikawa, F. Kagawa, and Y. Tokura, Phys. Rev. B **91**, 041122(R) (2015).
4. Chapter 6: "Band-filling dependence of thermoelectric properties in B20-type CoGe", N. Kanazawa, Y. Onose, Y. Shiomi, S. Ishiwata, and Y. Tokura, Appl. Phys. Lett. **100**, 093902 (2012).

Supervisor's Foreword

Continuous research over half a century on so-called B20-type compounds provides us with versatile stimulating physical properties. These are exemplified by non-collinear spin order, a non-trivial insulating state possibly due to the Kondo effect, and Dirac band dispersion, all of which originate from their chiral crystal structure. In particular, recent discovery of the novel spin structure "skyrmion" in B20-type magnets has brought a breakthrough in the emerging concept of topology in condensed matter physics. The skyrmion is a topological magnetic particle in real space. In its internal structure, constituent spins point in all the directions wrapping a sphere, which is mathematically described by a distinct topological index and can be a source of various emergent phenomena such as the topological Hall effect, ultralow current drive of skyrmions, and so on. Owing also to great progress in developing novel electronic states utilizing topology in the momentum space, such as graphene and topological insulators, a sweeping trend is to rewrite and refine various physical phenomena with a view of electronic topology, which is now leading to establishment of "emergent electromagnetism".

This thesis by Naoya Kanazawa presents his systematic study on a series of B20-type germanium compounds from a multidisciplinary approach. Many experimental methods used in the study, including high-pressure synthesis, molecular beam epitaxy, neutron diffraction, thermoelectric measurements, first-principles band calculations, and so on, offer new insights into both basic and applied physics. Three major topics are in noteworthy progress: formation of three-dimensional skyrmion crystal, electrical detection of skyrmions in nanocircuits, and enhancement of thermoelectric performance in the presence of Dirac electrons. The present study offers a guiding principle in utilizing topological aspects of electronic and magnetic structure to produce novel physical phenomena.

Tokyo
March 2015

Prof. Yoshinori Tokura

Acknowledgments

I would like to express my sincerest gratitude to Prof. Yoshinori Tokura for his continuing research guidance and encouragement in the course of this work.

I also appreciate the assistance of the following people: Prof. T. Arima (neutron experiments and valuable guidance), Prof. R. Arita (discussion), Dr. N. Egetenmeyer (SANS), Prof. Y. Endoh (discussion), Prof. J.L. Gavilano (SANS), Dr. T. Hara (TEM), Mr. T. Ideue (discussion), Prof. M. Ichikawa (thin film and valuable guidance), Prof. D.S. Inosov (SANS), Prof. S. Ishiwata (high-pressure synthesis), Prof. F. Kagawa (transport measurements and valuable guidance), Prof. K. Kakurai (neutron diffraction), Prof. M. Kawasaki (thin film and valuable guidance), Prof. B. Keimer (SANS), Dr. J.-H. Kim (SANS), Prof. K. Kimoto (TEM), Dr. M. Kubota (device fabrication), Dr. Y. Li (thin film), Prof. Y. Matsui (TEM), Dr. D. Morikawa (TEM), Prof. N. Nagaosa (Theoretical calculations and valuable guidance), Prof. K. Nakamura (discussion), Prof. K. Ohoyama (neutron diffraction), Dr. D. Okuyama (neutron diffraction), Prof. Y. Onose (discussion), Mr. K. Shibata (TEM), Dr. Y. Shiomi (thermal transport measurements), Prof. A. Tsukazaki (thin film and valuable guidance), Dr. S. Wakimoto (neutron diffraction), Dr. J.S. White (SANS), Mr. T. Yokouchi (thin film), and Dr. X.Z. Yu (TEM).

I am grateful to Profs. Naoto Nagaosa, Masashi Kawasaki, Daichi Chiba, and Taka-hisa Arima for valuable comments and recommendations on this thesis. Also, I thank all the members of the Tokura group for their great help in diverse ways. Finally, I would like to express my special gratitude to my parents for their support and encouragement for years.

Contents

Chapter 1
Introduction

1.1 Basic Informations on B20-type Compounds

Compounds of transition-metal element and group-14 element with composition ratio of 1:1 sometimes crystallize into a so-called B20-type structure. $3d$-transition-metal monosilicides have especially provided the area of condensed matter physics with several important phenomena. Helical spin order in MnSi was found in the initial stage of research on non-collinear spin structure [1]. MnSi is also one of prototypical examples of spin-fluctuation theory [2]. FeSi was predicted to be the first example of Kondo insulator of $3d$-transition-metal compounds [3]. CoSi is now a famous thermoelectric material showing large electrical conductivity and large thermopower simultaneously [4]. Here we introduce basic informations on B20-type compounds, such as the crystal structure and variety of compounds.

The crystal structure of the B20-type compound belongs to space group of $P2_13$, which is No. 198 in the international table. Pearson symbol and Strukturbericht symbol are also assigned as cP8 and B20, respectively. This structure lacks space inversion symmetry and is a chiral structure.

The unit cell of the B20-type compound includes four transition-metal atoms and four group-14 atoms, whose relative position in the unit cell is described as $(x, x, x; x+\frac{1}{2}, \frac{1}{2}-x, \bar{x}; \circlearrowright)$. For example, the parameters of FeSi are $x_{\mathrm{Fe}} = 0.1370 \pm 0.0020$ and $x_{\mathrm{Si}} = 0.842 \pm 0.004$ [5]. Above definitions are summarized in Table 1.1.

A schematic illustration of the cubic unit cell of the B20-type crystal is shown in Fig. 1.1. The lattice constant of FeSi, for example, is $a = b = c = 4.489$ Å [5]. The chiral structure is well exhibited in views from [111] crystal axis (lower panels in Fig. 1.1). Both the left- and right-handed atomic arrangements are possible and they cannot be superimposable with each other; the atoms stack like a spiral staircase with different rotation directions. This chiral-lattice structure influences the magnetic structure as discussed in Sect. 1.2.

Table 1.2 presents a list of space groups which compounds of transition metal (A) and group 14 element (B) belong to. The label of space group $P2_13$ (B20-type

N. Kanazawa, *Charge and Heat Transport Phenomena in Electronic and Spin Structures in B20-type Compounds*, Springer Theses,
DOI 10.1007/978-4-431-55660-2_1

Table 1.1 Definitions of the B20-type structure

Prototype	FeSi
Pearson Symbol	cP8
Strukturbericht Symbol	B20
Space group	$P2_13$
International table	198
Number of atoms in the unit cell	Fe: 4 atoms, Si: 4 atoms
Primitive translation vector	$(a, 0, 0; \circlearrowright)$
Relative position	$(x, x, x; x+\frac{1}{2}, \frac{1}{2}-x, \bar{x}; \circlearrowright)^{a}$

[a] Individual x values for Fe and Si

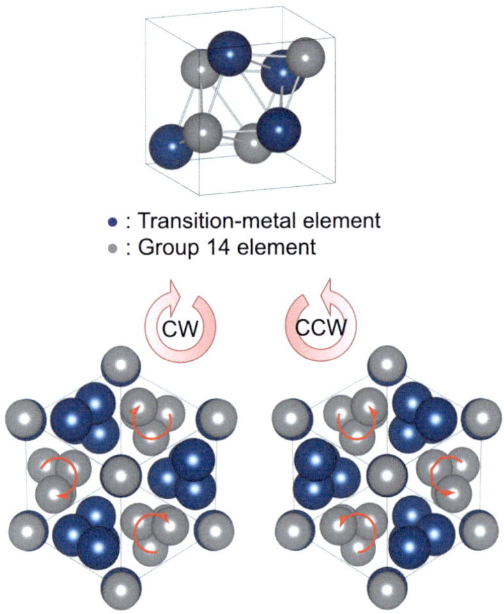

Fig. 1.1 Crystal structure of the B20-type compound. B20-type crystal structure is a chiral structure: Both left- and right-handed structures are present. Lower figures are views of both the structures from [111] crystal axis

structure) appears in the columns of 3d-elements (Cr, Mn, Fe, Co), 4d-elements (Ru, Rh), and 5d-elements (Hf, Re, Os). Here we note that Table 1.2 is based on the reports using crystal-growth methods at ambient pressure. High-pressure synthesis methods can sometimes transform AB compounds with different crystal structures to the B20-type because B20-type structure is the second closest packing among the crystal structures of AB [6]. A list of Strukturbericht Symbol in order of close packing is as follows: B2 ($Pm\bar{3}m$, CsCl) → B20 ($P2_13$, FeSi) → ($C2/m$, CoGe) → B35 ($P6/mmm$, CoSn). Compounds with less dense structure than B20-type one, such as CoGe-type and CoSn-type, have a chance to be transformed into B20-type by utilizing a high-pressure synthesis method. Indeed, we have succeeded in fabricating a series

Table 1.2 Space groups to which crystal structures of AB compounds belong

3d	Sc	Ti	V	Cr	Mn	Fe	Co	Ni	Cu	Zn
C	n/a	$Fm\bar{3}m$	$Fm\bar{3}m$	n/a	n/a	n/a	n/a	n/a	n/a	n/a
Si	$Cmcm$	$Pnm2$ $Pnma$	n/a	$P2_13$	$P2_13$	$P2_13$	$P2_13$	$Pnma$	n/a	n/a
Ge	$Cmcm$	n/a	n/a	$P2_13$	n/a $P2_13$[a]	$c2/m$ $P6/mmm$ $P2_13$	$C2/m$ $P2_13$[a]	$Pnma$	n/a	n/a
Sn	n/a	n/a	n/a	n/a	n/a	$P6/mmm$	$P6/mmm$	n/a	n/a	n/a
Pb	n/a	n/a	n/a	n/a	n/a	n/a	n/a	n/a	n/a	n/a
4d	Y	Zr	Nb	Mo	Tc	Ru	Rh	Pd	Ag	Cd
C	n/a	$Fm\bar{3}m$	$Fm\bar{3}m$	n/a	n/a	n/a	n/a	n/a	n/a	n/a
Si	$Cmcm$	$Cmcm$ $Pnma$	n/a	n/a	n/a	$P2_13$	$P2_13$ $Pnma$	$Pnma$	n/a	n/a
Ge	$Cmcm$	$Pnma$	n/a	n/a	n/a	$P2_13$	$Pnma$	$Pnma$	n/a	n/a
Sn	n/a	n/a	n/a	n/a	n/a	n/a	$P2_13$	$Pnma$	n/a	n/a
Pb	n/a	n/a	n/a	n/a	n/a	n/a	$P6/mmm$	$P\bar{1}$	n/a	$P4/mmm$

(continued)

Table 1.2 (continued)

5d		Hf	Ta	W	Re	Os	Ir	Pt	Au	Hg
C		n/a	$Fm\bar{3}m$	$P\bar{6}m2$	n/a	n/a	n/a	n/a	n/a	n/a
Si		$Pnma$	n/a	n/a	$P2_13$	$P2_13$	$Pnma$	$Pnma$	n/a	n/a
Ge		$Pnma$	n/a	n/a	n/a	n/a	n/a	$Pnma$	n/a	n/a
Sn		$P2_13$	n/a	n/a	n/a	n/a	n/a	$P6_3/mmc$	$P6_3/mmc$	n/a
Pb	n/a	n/a	n/a	n/a	n/a	n/a	n/a	n/a	n/a	n/a

Here, A = transition-metal element and B = group-14 element. Space group $P2_13$ represents B20-type structure. If an aimed composition does not show B20-type crystal structure and is in the neighborhood of the compounds with B20-type structure in this periodic table, high-pressure synthesis may be effective for transformation of the crystal structure into the B20-type. (e.g., MnGe)

[a]High-pressure phase

of B20-type germanium compounds (MnGe, $Mn_{1-x}Fe_xGe$, FeGe, $Fe_{1-x}Co_xGe$, CoGe) by using a cubic-anvil type high-pressure apparatus. Detail procedure of crystal growth method is described in Sect. 2.1.

1.2 Helical and Skyrmion Spin Structure

1.2.1 Overview

There are three well-separated energy scales in magnetic interactions interplaying in the B20 magnets: ferromagnetic exchange interaction, Dzyaloshinsky-Moriya (DM) interaction, and crystalline field interactions (magnetic anisotropy energy) in decreasing order of strength. The hierarchy of the magnetic interactions results in a ubiquitous magnetic properties and a generic theoretical description of B20 magnets as discussed in detail below.

The ferromagnetic interaction ($\propto \mathbf{S}_i \cdot \mathbf{S}_j$) works as a driving force for a parallel order of spins, while the DM interaction ($\propto \mathbf{S}_i \times \mathbf{S}_j$) for a spin arrangement with 90° mutual spin angles. Due to the competition between the ferromagnetic and the DM interactions, the ground state of magnetic order is a long-period spiral spin structure. Its modulation direction is pinned along a certain crystal axis by the magnetic anisotropy.

Magnetic phase diagrams of B20 magnets share a universal profile. The phase diagram for MnSi is exemplified in Fig. 1.2 with schematic illustrations of observed magnetic structures. At low magnetic fields: a helical structure with its wave vector $\mathbf{Q}$ parallel to equivalent $\langle 111 \rangle$ crystal axes (Fig. 1.2c). At intermediate fields: a conical structure running along the applied magnetic field (Fig. 1.2d). At high magnetic fields: ferromagnetic state. In a small pocket just below the transition temperature T_N: a skyrmion lattice stretching along magnetic field direction (Fig. 1.2b).

1.2.2 Helical Structure

A spiral spin ordering, helical structure, was observed in MnSi at the early stage of research on non-collinear magnetic structure [1]. Figure 1.3 shows small-angle neutron scattering (SANS) patterns on MnSi [1]. The peak intensities in Fig. 1.3a, b represent that a periodically-modulated magnetic structure forms along equivalent $\langle 111 \rangle$ crystal axes at low magnetic fields. That magnetic structure was confirmed to be a helical spin structure by Lorentz transmission electron microscopy (LTEM) observations [8] and theoretical calculations [9–11]. With application of larger magnetic fields, the helical structure transforms into a conical structure with modulation vector $\mathbf{Q}$ flopped along the field direction (Fig. 1.3c, d), and finally spin-collinear (ferromagnetic) state above the critical field H_c.

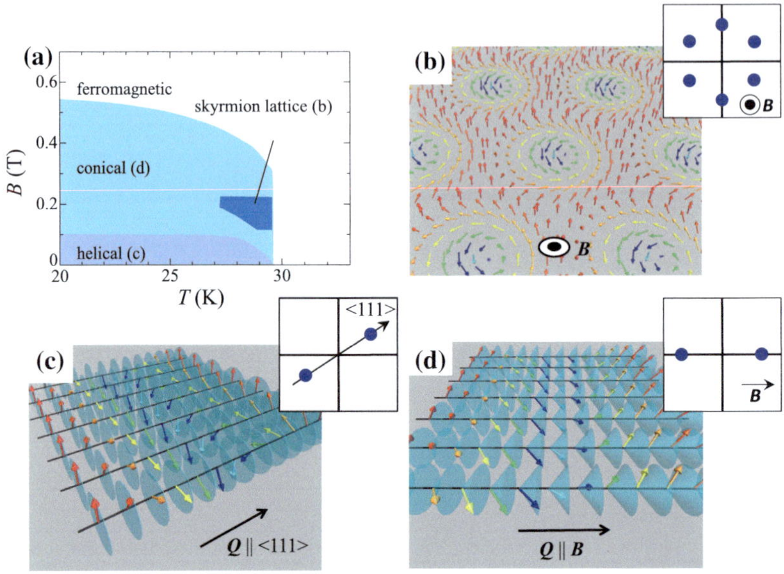

Fig. 1.2 **a** A typical phase diagram of a B20-type magnet MnSi in the T-B plane [7]. The magnetic structure in each phase is illustrated in panels (**b**)–(**d**): **b** skyrmion lattice, **c** helical structure, **d** conical structure

The flipping process of the modulation vector appears in magnetic field (H) dependence of magnetization (M), which is also confirmed by SANS experiments [12–14]. Figure 1.4 presents magnetic field dependences of magnetization and SANS patterns [13] with schematic illustrations of spin states in $Fe_{1-x}Co_xSi$ specimens. In the course of ramping magnetic field up to the critical magnetic field, weakly pinned (almost randomly distributed) **Q** at low magnetic fields flips toward the magnetic field direction, which leads to the stepwise increase of magnetization. In the course of decreasing magnetic field back to zero, **Q** remains parallel to the magnetic field direction, which results in the smooth change in the M-H curve.

The formation of the helical structure has been completely determined by LTEM observation [8]. Figure 1.5 shows typical LTEM images of helical magnetic structure forming in B20-type $Fe_{0.5}Co_{0.5}Si$. A stripe pattern in Fig. 1.5b is observed below the transition temperature $T_N = 38$ K, while no contrast pattern in Fig. 1.5a is detected above T_N. A transport-intensity-equation (TIE) analysis [15], which can estimate in-plane magnetization, confirms a formation of the helical magnetic structure (so-called proper screw type), whose spin rotation plane is perpendicular to the modulation direction (see also Fig. 1.2c). The helical period λ is evaluated as 90 nm from the captured contrast image (Fig. 1.4d), which is consistent with a neutron diffraction experiment [16].

Theoretical studies immediately following the neutron diffraction experiments also demonstrated that the periodic magnetic structure in B20-type magnets is indeed

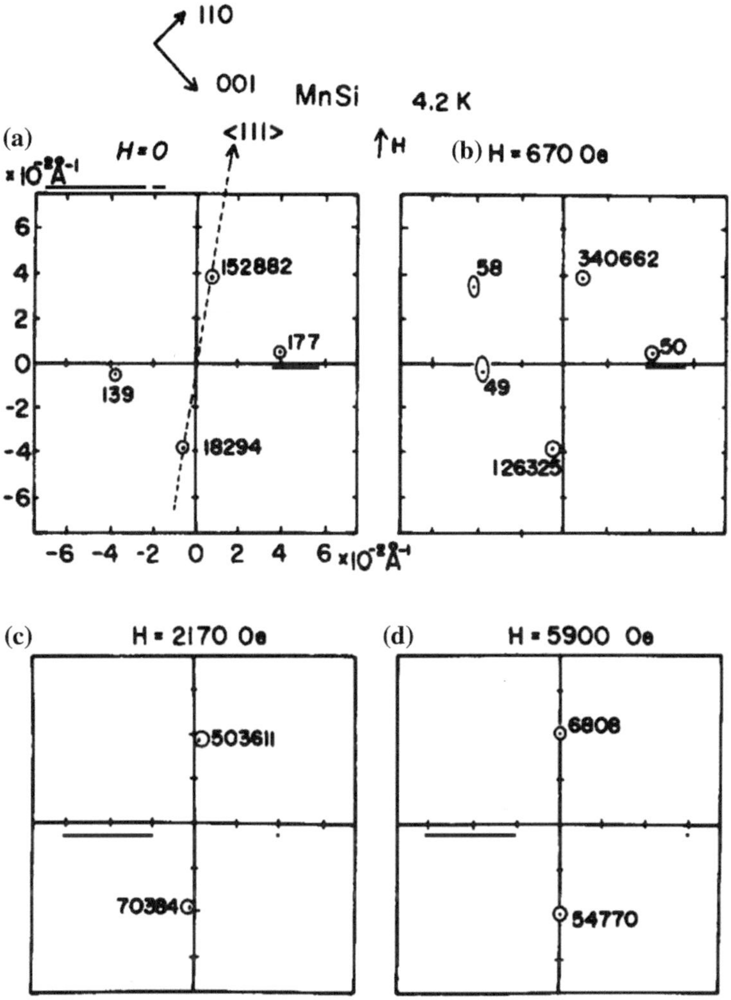

Fig. 1.3 The first observation of helical ordering of spins in B20-type MnSi by SANS [1]. The modulation of the helical structure is pinned along equivalent ⟨111⟩ crystal axes at low magnetic fields in panels (**a**) and (**b**), while it flops along the applied large magnetic fields in panels (**c**) and (**d**). Reprinted from Ref. [1], Copyright 1976, with permission from Elsevier

the helical magnetic structure [9–11]. Taking into account the space group of B20-type crystal $P2_13$, the free energy F and the Hamiltonian H of the spin system, which are functionals of the spatial variation of the spins $\mathbf{S}(\mathbf{r})$, are expressed respectively as follows [11]:

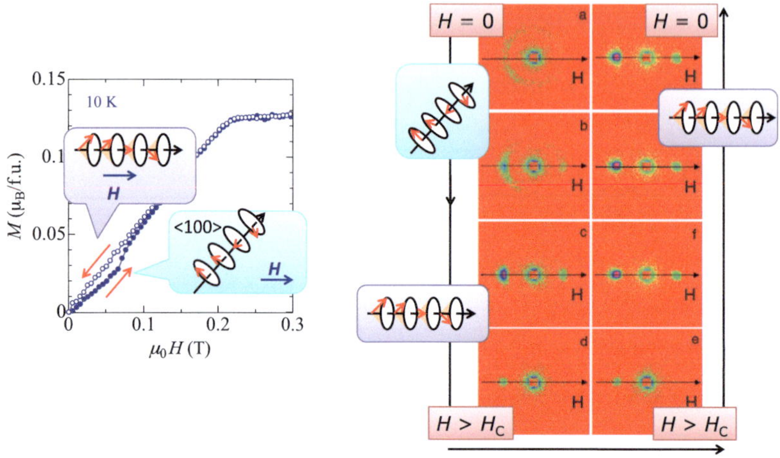

Fig. 1.4 Schematic illustrations of the flopping process of the helical modulation in a magnetic field development, which is compared with magnetization data (*left panel*) and SANS patterns (*right panel*) [13]. Reprinted with permission from Ref. [13]. Copyright 2007 by American Physical Society

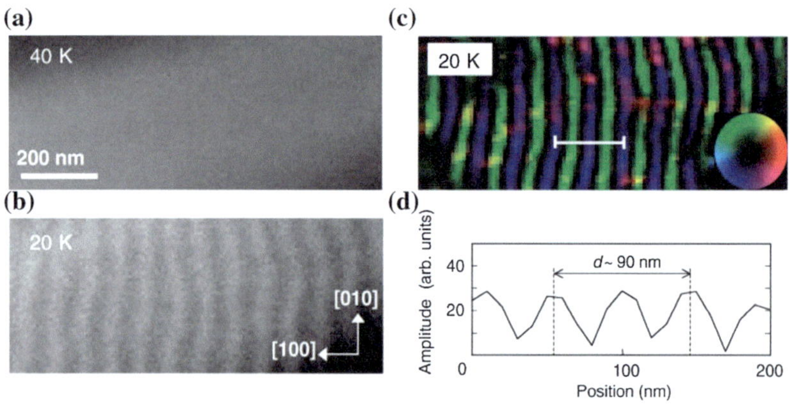

Fig. 1.5 Real-space observation of helical magnetic structure in B20-type $Fe_{0.5}Co_{0.5}Si$ by Lorentz transmission electron microscopy [8]. Below the transition temperature ($T_N = 38$ K), a contrast of stripe pattern is observed in panel (**b**). Transport-Intensity-Equation (TIE) analysis confirms the spin ordering belongs to the proper screw type as shown in panel (**c**). From [8]. Reprinted with permission from AAAS

$$
\begin{aligned}
F = & \left[\frac{1}{2}(\chi(0)^{-1} + JQ^2)\langle S^2\rangle_p + \frac{1}{4}\gamma\langle S^4\rangle_p\right] \\
& + D[\mathbf{S}_1 \times \mathbf{S}_2]\cdot\mathbf{Q} \\
& + \frac{1}{2}[A_1(\langle 3S_z^2 - S^2\rangle_p(3Q_z^2 - Q^2) + 3\langle S_x^2 - S_y^2\rangle_p(Q_x^2 - Q_y^2))
\end{aligned}
$$

$$+ A_2(\langle 3S_z^2 - S^2\rangle_p(Q_x^2 - Q_y^2) - \langle S_x^2 - S_y^2\rangle_p(3Q_z^2 - Q^2))$$
$$+ 4A_3(\langle S_xS_y\rangle_p Q_xQ_y + \langle S_yS_z\rangle_p Q_yQ_z + \langle S_zS_x\rangle_p Q_zQ_x)]$$
$$- \frac{1}{2}K_1(\langle S_x^4\rangle_p + \langle S_y^4\rangle_p + \langle S_z^4\rangle_p),$$

$$H = \int dr \left[\frac{J}{2}\sum_{ii'}\left(\frac{\partial S_i(r)}{\partial r_{i'}}\right)^2 - DS(r)\cdot(\nabla \times S(r)) - \frac{1}{2}K_1\sum_i S_i(r)^4 - H \cdot S(r)\right]$$

As the neutron diffraction experiments clearly demonstrated that the modulated spin structure has a single $\mathbf{Q}$, we employ the function $\mathbf{S}(\mathbf{r})$ in the form of $\mathbf{S}(\mathbf{r}) = \mathbf{S}_1\cos(\mathbf{Q}\cdot\mathbf{r}) + \mathbf{S}_2\sin(\mathbf{Q}\cdot\mathbf{r})$. Here we also note that the magnitude of modulation vector Q is small and we take the terms up to the forth and the second orders of S and Q. In the description of the free energy, the first term represents the exchange interaction, the second is Dzyaloshinsky-Moriya (DM) type interaction, which vanishes in the system with inversion symmetry. This DM interaction $D[\mathbf{S}_1\times\mathbf{S}_2]\cdot\mathbf{Q}$ is important for the stabilization of the long period helical spin order with the helical plane perpendicular to $\mathbf{Q}$. The DM interaction minimize the free energy at $Q \approx |D|/J$ and its sign determines the rotation direction (helicity) of the spiral structure. The third and the forth terms are anisotropy energies dependent on S and Q and independent of Q, respectively, which determine the pinned direction of the spin modulation. The first, second, and third terms of Hamiltonian are the exchange interaction, DM interaction, and anisotropic interaction, respectively.

Comparing free energies of different magnetic structures in the form of $\mathbf{S}(\mathbf{r}) = \mathbf{S}_1\cos(\mathbf{Q}\cdot\mathbf{r}) + \mathbf{S}_2\sin(\mathbf{Q}\cdot\mathbf{r})$, the helical magnetic structure shows the minimum free energy in comparison to other structures, such as a sinusoidal spin structure, the ferromagnetic structure, and so on, when the anisotropy energy is sufficiently small.

Following relations are also derived as consequences of the above Hamiltonian:

$$Ja^2 = O(k_BT_N), \tag{1.1}$$
$$Q \approx D/J, \tag{1.2}$$
$$H_c \approx D^2S/J. \tag{1.3}$$

The helical period can be roughly estimated by substituting the measures of magnetization into the above relations, which is also discussed in Sect. 3.1.

Formation of helical structure is ubiquitously observed in B20-type magnets except for MnGe [17, 18], in which a skyrmion-lattice state is possibly the ground state as discussed in Chap. 4. Transition temperatures T_N and helical periods λ of B20-type magnets are listed in Table 1.3 for reference. The helical period ranges roughly from 3 nm to 230 nm.

1.2.3 Skyrmion

In addition to the helical and the conical structures, another magnetic phase exists just below T_N, which is called as A-phase [26–28]. The magnetic structure in the A-phase

Table 1.3 List of transition temperatures (T_N) and helical periods (λ) of helimagnets with B20-type crystal structure

Material		T_N (K)	λ (nm)	References
MnSi	Bulk	30	18	Ref. [1]
	Epitaxial thin film	45	8.5	Ref. [19]
$Mn_{1-x}Fe_xSi$	$x = 0.06$	16.5	12.5	Ref. [20]
	$x = 0.08$	10.6	11	Ref. [20]
	$x = 0.10$	6.8	10	Ref. [20]
$Fe_{1-x}Co_xSi$	$x = 0.10$	11	43	Refs. [16, 21]
	$x = 0.5$	36	90	Refs. [16, 21]
	$x = 0.6$	24	174	Refs. [16, 21]
	$x = 0.7$	7	230	Refs. [16, 21]
MnGe	$T = 20$ K	170	3	Ref. [17]
	$T = 100$ K		3.4	Ref. [17]
	$T = 150$ K		5.5	Ref. [17]
$Mn_{1-x}Fe_xGe$	$x = 0.35$	150	4.7	Ref. [22]
	$x = 0.5$	185	14.5	Ref. [22]
	$x = 0.7$	210	77	Ref. [22]
	$x = 0.84$	220	220	Ref. [22]
FeGe	Bulk	278	70	Ref. [23]
Cu_2OSeO_3	Bulk	59	62	Ref. [24]
	Thinned plate		50	Ref. [25]

had yet to be elucidated until a SANS experiment on MnSi by Mühlbauer et al. [7]. The SANS experiment was conducted in the condition that magnetic field is applied parallel to the incident neutron beam. Figure 1.6 shows a sketch of the experimental setup and SANS patterns. In the T-H region of the A-phase, six magnetic Bragg reflections are observed on a regular hexagon which are perpendicular to the magnetic field (Fig. 1.6c–e). This means the magnetic structure has independent three wave vectors, because one magnetic modulation $\mathbf{Q}$ diffracts the unpolarized neutron beam to two positions (Q and $-Q$) in the reciprocal space. Mühlbauer et al. suggested that the realized magnetic structure is a superposition of three helical structures corresponding to three independent wave vectors with mutual angles of 120° [7]. This multiple-Q magnetic structure can be regarded as a triangular lattice of spin-swirling objects, i.e., a skyrmion lattice (see Figs. 1.7 and 1.8). Here it is particularly worth noting that the skyrmion formation in noncentrosymmetric magnets was theoretically predicted by Bogdanov et al. [29, 30] before the experimental breakthroughs [7, 31].

The "skyrmion" was originally a model describing baryons existing in the continuous field, which was invented by the nuclear physicist Tony Skyrme [32]. Recently the concept of the skyrmion is variously used if structures fulfills the following condition: When we map an order parameter (e.g. spin) of an object (e.g. skyrmion) in a coordinate space (e.g. real space) from the coordinate variable (e.g. position) to

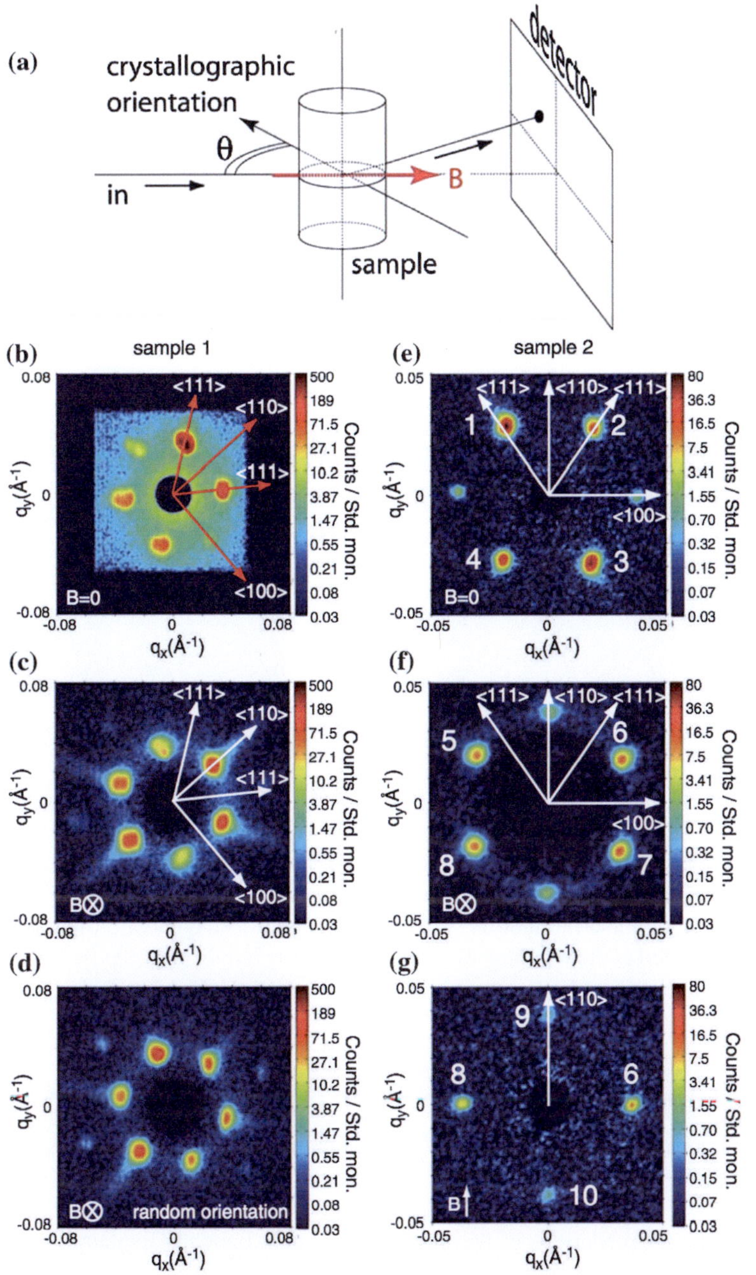

Fig. 1.6 Experimental setup of the SANS on MnSi and variation of SANS patterns with magnetic field [7]. Six-fold peak intensities in panels (**b**)–(**e**) show that the spins are arranged in a triangular lattice form in the plane perpendicular to the applied magnetic field. This is the first observation of skyrmion-lattice state by SANS. From [7]. Reprinted with permission from AAAS

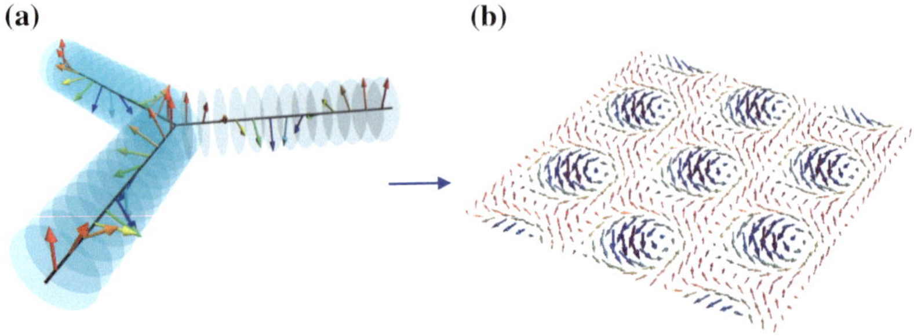

Fig. 1.7 Schematic illustrations of **a** superposition of three helical structures with 120° mutual angles and **b** the consequent spin structure: the skyrmion lattice

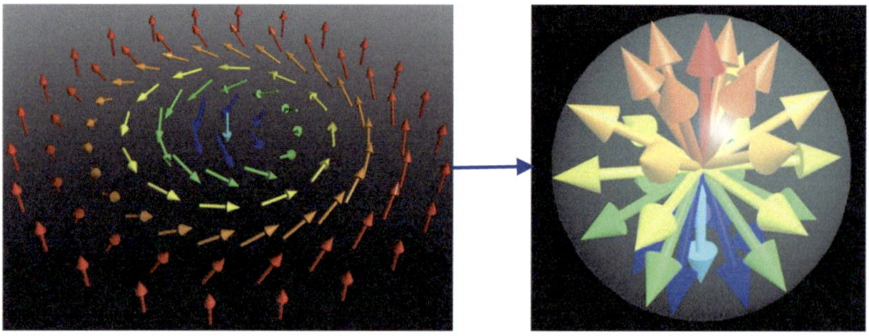

Fig. 1.8 Skyrmion spin structure and its projection onto the spherical spin space. The directions of the spins in one skyrmion wraps the sphere

the rotational parameter (e.g. spin direction) of the order-parameter space (e.g. spin space), the number of the order parameter's rotation in the order-parameter space, so-called topological number, is integer. In the field of magnetism, the topological number is called as the winding number. The winding number (the skyrmion number) in the target manifold [2 dimensional (2D) space] can be calculated as

$$w_\mu = \int \frac{1}{8\pi} \varepsilon_{\mu\nu\lambda} \hat{\mathbf{n}} \cdot \left(\frac{\partial \hat{\mathbf{n}}}{\partial \nu} \times \frac{\partial \hat{\mathbf{n}}}{\partial \lambda} \right) \mathrm{d}\nu \mathrm{d}\lambda. \tag{1.4}$$

If the skyrmion magnetic structure in B20 compounds is mapped to the spin space, the direction of the spins wraps a sphere exactly once and the winding number is $\int \frac{1}{4\pi} \hat{\mathbf{n}} \cdot (\partial_x \hat{\mathbf{n}} \times \partial_y \hat{\mathbf{n}}) \mathrm{d}x \mathrm{d}y = -1$ (see Fig. 1.8).

Mean-field calculations of the phenomenological Landau-Ginzburg model with Gaussian fluctuations support the suggested multiple-Q structure and well reproduce the magnetic phase diagram (Fig. 1.9) [7]. Along with the full calculation, an intuitive explanation for the skyrmion lattice formation is also demonstrated in Ref. [7]. A functional of the free energy is described as:

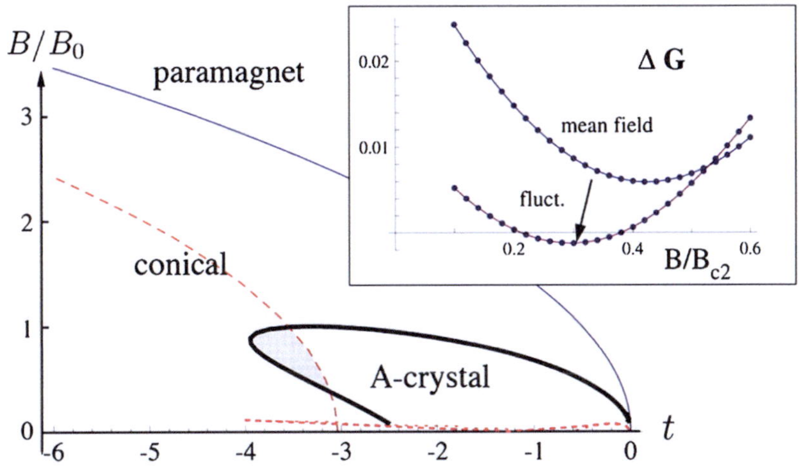

Fig. 1.9 Theoretical calculation of phase diagram in the bulk B20-type magnets with using a model which attributes the three mode coupling to the origin of the skyrmion-lattice formation. From [7]. Reprinted with permission from AAAS

$$F[\mathbf{M}] = \int d^3r\ [r_0\mathbf{M}^2 + J(\nabla\mathbf{M})^2] + 2D\mathbf{M}\cdot(\nabla\times\mathbf{M}) + U\mathbf{M}^4 - \mathbf{B}\cdot\mathbf{M}], \quad (1.5)$$

where $\mathbf{B}$ is the external magnetic field and r_0, J, D, U are parameters ($J, U > 0$). The real order parameter, which describes the spacial variation of magnetization, should exclude the ferromagnetic component induced by the magnetic field: $\Phi = \mathbf{M} - \mathbf{M}_f$. Here $\mathbf{M}_f$ is the ferromagnetic component $\mathbf{M}_f = \int \mathbf{M}d^3r/V$. The terms of the exchange interaction and the DM interaction (the quadratic part) stabilize a helical structure denoted by $m[\mathbf{n}_1\cos(\mathbf{Q}\cdot\mathbf{r}) + \mathbf{n}_2\cos(\mathbf{Q}\cdot\mathbf{r})]$, where $\mathbf{n}_1$ and $\mathbf{n}_2$ are orthogonal to each other. Limiting to the quadratic order, arbitrary linear combinations of such helices minimize the free energy. The term of $\mathbf{M}^4$ lifts the degeneracy and determines the magnetic structure. The term $\mathbf{M}^4$ contains $\Phi^2\Phi\cdot\mathbf{M}_f$, whose Fourier-transformed expression is

$$\int d^3r\Phi^2\Phi\cdot\mathbf{M}_f = \sum_{\mathbf{q}_1,\mathbf{q}_2,\mathbf{q}_3\neq 0} (\mathbf{M}_f\cdot\mathbf{m}_{\mathbf{q}_1})(\mathbf{m}_{\mathbf{q}_2}\cdot\mathbf{m}_{\mathbf{q}_3})\delta(\mathbf{q}_3+\mathbf{q}_3+\mathbf{q}_3). \quad (1.6)$$

This term can reduce the free energy when a combination of wave vectors fulfills $\mathbf{Q}_1 + \mathbf{Q}_2 + \mathbf{Q}_3 = 0$, which leads the stabilization of skyrmion lattice state. In fact, the skyrmion lattice formation can become the ground state by introducing thermal fluctuation.

Although the SANS experiments and the theoretical calculations strongly suggest the existence of the skyrmion lattice, non-trivial problems remain: Phases ϕ of the superposed helices $m_i[\mathbf{n}_1\cos(\mathbf{Q}_i\cdot\mathbf{r}+\phi_i)+\mathbf{n}_2\cos(\mathbf{Q}_i\cdot\mathbf{r}+\phi_i)]$ cannot be determined.

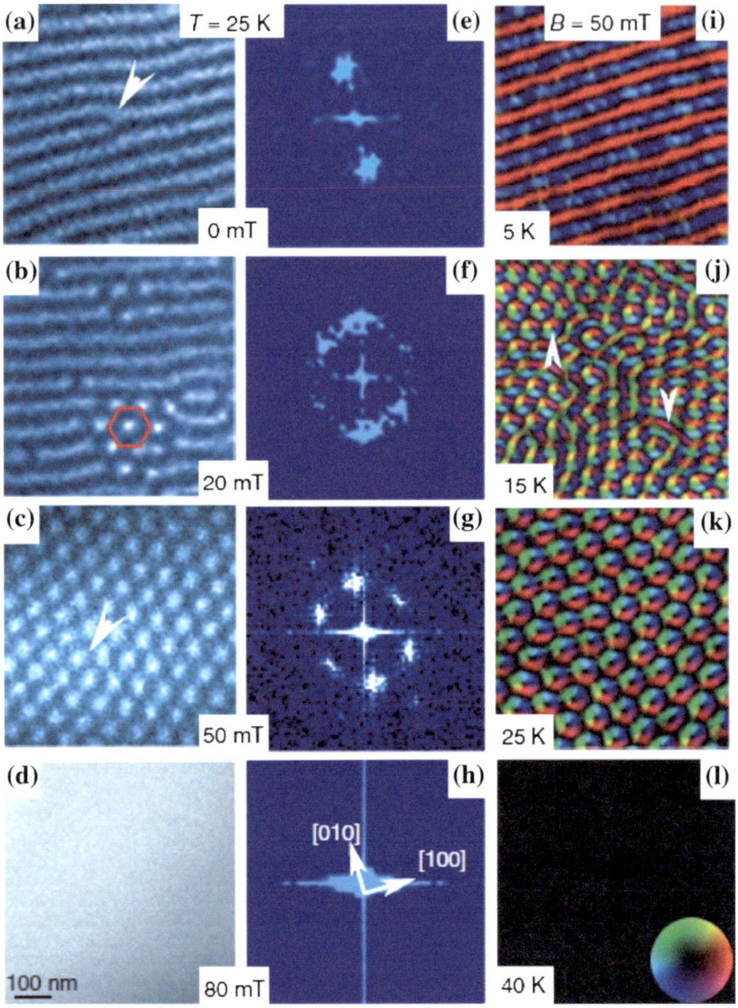

Fig. 1.10 Real-space observation of skyrmion spin structures by Lorentz transmission electron microscopy [31]. Defocused images [panels (**a**)–(**d**)], Fourier transformed images [panels (**e**)–(**h**)], and TIE-analysis images [panels (**i**)–(**l**)] are presented. Transformation of spin structure from helical order to coexistence of helical structure and skyrmions, skyrmion-lattice state, and ferromagnetic state is observed. This report provides the first and definite proof of emergence of skyrmion spin structure. Reprinted by permission from Macmillan Publishers Ltd: Ref. [31], copyright 2010

Furthermore, a multi-domain state of helical structures can also produce the six-fold diffraction intensities.

Real-space observation by LTEM method successfully demonstrates the spin arrangement in $Fe_{0.5}Co_{0.5}Si$ [31]. Here we note that the specimens were thinned to be below 100 nm thick for electron transmission, and thus the thickness is comparable to the helical period of $Fe_{0.5}Co_{0.5}Si$ (90 nm). Figure 1.10 shows defocused TEM images, their Fourier transformed patterns, and TIE-analysis images of $Fe_{0.5}Co_{0.5}Si$, the magnetic field is applied perpendicular to the plane. Circle-shaped patterns appear in the back ground of stripe pattern in Fig. 1.10b and are aligned in the form of triangular lattice in Fig. 1.10c. The maps of lateral magnetization distribution (TIE images in Fig. 1.10) reveals that the magnetization swirls in the same rotation direction in each circle contrast.

Other two important findings were shown by the LTEM observation on the thinned $Fe_{0.5}Co_{0.5}Si$: particle behavior of skyrmions and a large extension of skymion phase. Those expand significantly the possibilities of technological applications to memory devices using skyrmions as information bits.

Particle behavior of skyrmions can be seen in the variation of the magnetic texture with changing magnetic field (Fig. 1.11a–c). Skyrmion structures coexist with a helical structure (Fig. 1.11a) or a ferromagnetic structure (Fig. 1.11c) as well as a triangular lattice of skyrmions (Fig. 1.11b), which demonstrates that single skyrmions can exist as stable or metastable state. This particle behavior is beyond the prediction of the Landau-Ginzburg model introduced above.

The extended skyrmion phase is well presented in the T-B phase diagram, i.e., the contour map of skyrmion density (Fig. 1.11). Compared to the small skyrmion phase of bulk samples stabilized by thermal fluctuation (Fig. 1.2a), skyrmions exist in a wide T-B region denoted by H+Sk, SkX and FM+Sk including zero temperature (Fig. 1.11d). The observed phase diagram is reproduced by a Monte-Carlo simulation of a two-dimensional spin system with DM interaction (Fig. 1.11h). This indicates that the spin system in a thinned B20-magnet can be almost regarded as a 2D spin system if the sample thickness is of the order of the helical period. The helical and conical states with a modulation vector normal to the thinned plate seems unfavorable state in this "pseudo"-2D spin system (they are obviously impossible states in the genuine-2D spin system), and thus they give way to the skyrmion states. The helical

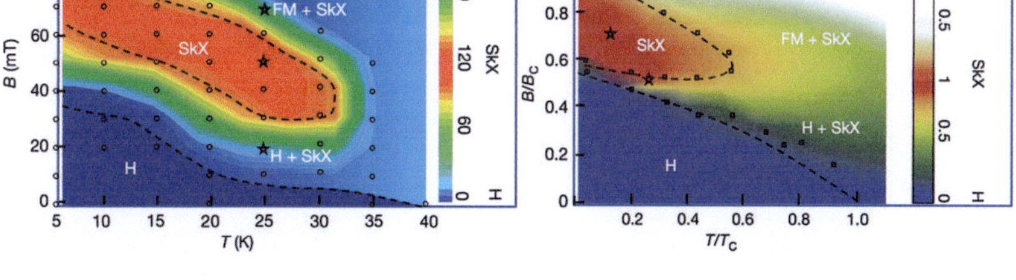

Fig. 1.11 Experimental and theoretical phase diagrams of B20-type $Fe_{0.5}Co_{0.5}Si$ [31]. Theoretical studies based on a two-dimensional model well reproduce the experimental T-H phase diagram. Reprinted by permission from Macmillan Publishers Ltd: Ref. [31], copyright 2010

or conical state with **Q** normal to the plate and parallel to the magnetic field may receive less energy gain from DM interaction due to the incomplete spiral structure in which the modulation runs less than one helical period.

1.2.4 Other Nontrivial Spin Structures

An additional magnetic phase of MnSi above a certain pressure ($p_c = 14.6$ kbar) expands possibilities of emergence of nontrivial magnetic textures in chiral-lattice magnets. Figure 1.12 shows a phase diagram of MnSi in the T-p plane with typical neutron diffraction patterns at ambient and high pressures [33]. Resistivity and magnetic susceptibility measurements trace the helical ordering temperature T_N (T_c in Fig. 1.12) as a function of pressure; the magnetic order and the ordered magnetic moment seem to vanish at p_c. Above p_c, however, a large magnetic scattering intensity still remains (only several tens of percent reduced from the ambient pressure value) and is located over a small sphere in the reciprocal space with broad maxima in the $\langle 110 \rangle$ directions. The magnetic structure is firstly interpreted as local scales long-range order in all three spatial directions rather than a short-range order, and thus it is referred to as "partial order" [33].

However, the observation of the clear diffraction peaks along $\langle 110 \rangle$ crystal axes attracts other theoretical interpretations, such as a square skyrmion lattice [34], a bcc spin crystal [35], and periodic arrangements of double-twist cylinders [36]. Here, we introduce a discussion using the Ginzburg-Landau free energy, which derives solutions of many types of multiple-Q structures or spin crystals, including the triangular skyrmion lattice [35].

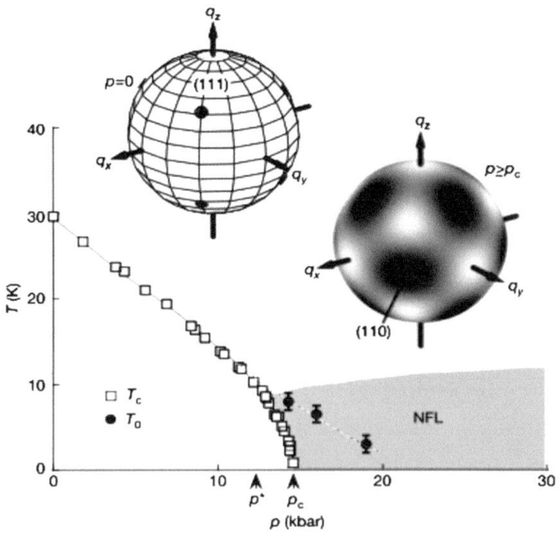

Fig. 1.12 Phase diagram in the pressure-temperature plane and typical neutron scattering patterns at an ambient pressure and at the high-pressure phase of the so-called partial-order state. Reprinted by permission from Macmillan Publishers Ltd: Ref. [33], copyright 2004

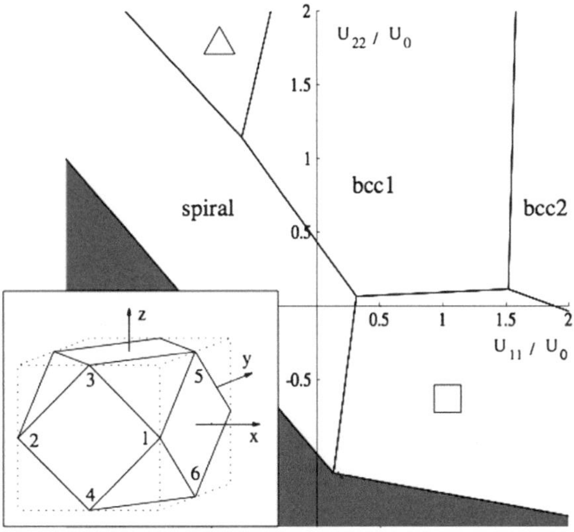

Fig. 1.13 Phase diagram of spin structures in the model of helical-mode coupling [35]. Various types of spin crystal states can be possible by changing anisotropy parameters. Reprinted with permission from Ref. [35]. Copyright 2006 by American Physical Society

To quadratic order of the free energy, a single spiral structure and linear combinations of several spin spirals (multiple-Q structures or spin crystals) are degenerate. The degeneracy can be lifted by the fourth-order term of $\mathbf{M}$:

$$F_4 = \sum_{\mathbf{q}_1,\mathbf{q}_2,\mathbf{q}_3} U(\mathbf{q}_1, \mathbf{q}_2, \mathbf{q}_3)(\mathbf{M}_{\mathbf{q}_1} \cdot \mathbf{M}_{\mathbf{q}_2})(\mathbf{M}_{\mathbf{q}_3} \cdot \mathbf{M}_{\mathbf{q}_4}), \tag{1.7}$$

where U is coupling function between different modes, $\mathbf{M}_\mathbf{q}$ is a Fourier mode of $\mathbf{M}(\mathbf{q})$, and $\mathbf{q}_4 = -(\mathbf{q}_1 + \mathbf{q}_2 + \mathbf{q}_3)$.

Phase diagram of realized spin state in the plane of coupling coefficients is presented in Fig. 1.13. Here the coefficients are defined as the followings: $U(\theta, \phi) = U_0 + U_{11} \sin\theta \cos\phi + U_{20}(3\cos^2\theta - 1) + U_{22} \sin^2\theta \cos 2\phi$, $2\theta = \arccos(\hat{\mathbf{q}}_1 \cdot \hat{\mathbf{q}}_2)$, and $\phi/2 = \arccos[(\hat{\mathbf{q}}_2 - \hat{\mathbf{q}}_1) \cdot \hat{\mathbf{q}}_3/(1 - \hat{\mathbf{q}}_1 \cdot \hat{\mathbf{q}}_2)]$. Several types of spin crystals appear with changes of coupling coefficients in addition to spin spiral state. The bcc crystals especially explain the neutron intensity peaks along $\langle 110 \rangle$ and possible candidates for the partial order. Furthermore this result demonstrates a rich variety of spin textures possibly emerging in B20-magnets.

1.3 Band Structure

To understand not only transport properties but also magnetic properties in B20-type compounds, it is important to know their band structure. Many band structures of B20-type compounds have been calculated (e.g. Refs. [37–39]), and it is widely recognized that the rigid band approximation works well for the B20-systems: The

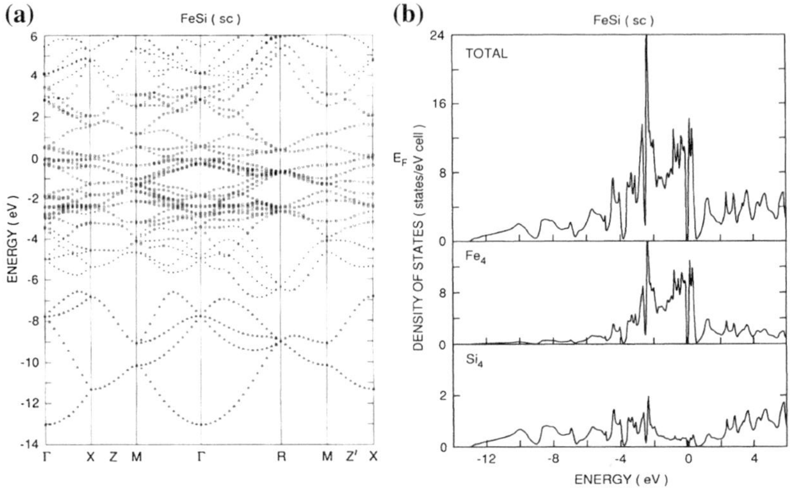

Fig. 1.14 **a** Band structure and **b** density of state of B20-type FeSi. Reprinted with permission from Ref. [38] Copyright 1993 by American Physical Society

Fermi energy E_F varies with the electron filling (nominal composition), while the energy-dispersion structure remains almost the same. We introduce in Fig. 1.14a band structure and density of states [DOS, $D(E)$] of FeSi [38]. When limiting the transition metal elements to Mn, Fe, and Co, which are the main focus of this thesis, E_F lies in the range between -3 eV and 3 eV, where the $3d$ character of transition metal is dominant (Fig. 1.14b).

1.3.1 Magnetic Ordering and Band Structure

The Stoner model, which employs itinerant description of electrons with an exchange energy, is applicable to the B20-type magnets. According to the Stoner criterion for ferromagnetism $JD(E_F) > 1$, the large DOS due to the narrow band structure (non-bonding character of the $3d$ electrons) around E_F is the origin of the magnetic ordering in B20-type compounds except for compositions around CoSi (E_F is located around 0.5 eV in the FeSi band structure, where a pseudogap is formed) (see also Fig. 1.14).

One-order higher transition temperatures in B20-germanides than B20-silicides can be understood in terms of their narrower band dispersion (the larger DOS). The distances between transition-metal atoms in B20-germanides are longer than those in B20-silicides because of the larger atomic radius of Ge than Si, which leads to the reduced transfer integral between transition-metal atoms in the germanium compounds, i.e., the narrower band dispersion.

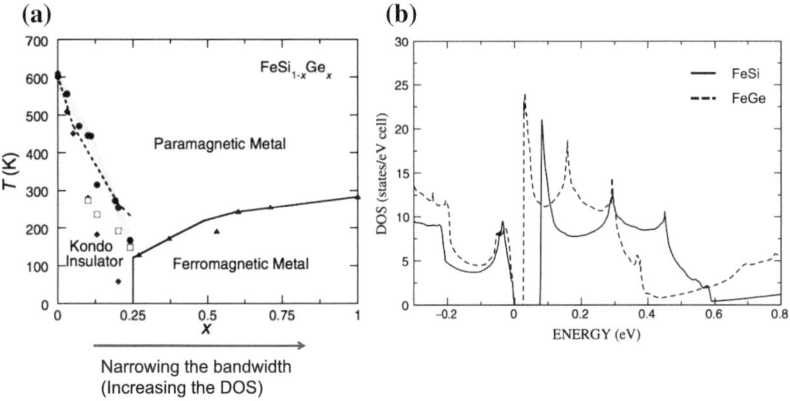

Fig. 1.15 **a** Phase diagram of physical properties in $FeSi_{1-x}Ge_x$ [40]. Insulator-metal transition occurs at $x = 0.25$ accompanied by ferromagnetic transition. **b** Density of state (DOS) of FeSi and FeGe [41]. The germanium compound has the larger DOS than the silicide. Reprinted with permission from Refs. [40, 41]. Copyright 2003 and 2002 by American Physical Society, respectively

Insulator-metal transition (IMT) in $FeSi_{1-x}Ge_x$ captures essential features of the above discussion (Fig. 1.15) [40, 41]. FeSi is a band insulator with a narrow band gap ($\approx$60 meV).[1] With decreasing the transfer integral (substituting Ge for Si) (Fig. 1.15b), the insulator-metal transition occurs at $x = 0.25$ simultaneously accompanied by magnetic ordering (Fig. 1.15a). It is unusual that narrowing the band width drives a transition from insulator to metal. LDA+U calculations [41] gave a full explanation that the IMT is induced by the paramagnet-ferromagnet transition, which can be intuitively understood in terms of the Stoner model.

1.3.2 Remark on Dirac Electrons in CoSi

Focusing on the calculated band structure of CoSi (e.g., see Ref. [39]), Dirac electrons, whose energy has a linear dependence on momentum, exist at the Fermi energy, which has not been referred to in previous reports. Dirac electrons in condensed matter can be (ideally) treated as zero mass particles like photons or neutrinos, and they obey the Dirac equation, not the Schrödinger equation. Dirac materials, such as graphene [42] and topological insulators [43], have been attracting much attention due to their characteristic properties. In this thesis, we focus on Dirac electrons in B20-compounds as the source of large Seebeck coefficient as discussed in Chap. 6.

[1] Many models for FeSi have been proposed, such as Kondo insulator [3].

References

1. Y. Ishikawa, K. Tajima, D. Bloch, M. Roth, Solid State Commun. **19**, 525 (1976)
2. T. Moriya, *Spin Fluctuations in Itinerant Electron Magnetism* (Springer, Berlin, 1985)
3. G. Aeppli, Z. Fisk, Comments. Condens. Matter Phys. **16**, 155 (1992)
4. D.M. Rowe, *CRC Handbook of Thermoelectrics* (CRC, Boca Raton, FL, 1995)
5. L. Pauling, A.M. Soldate, Acta Cryst. **1**, 212 (1948)
6. V.I. Larchev, S.V. Popova, J. Less-Common Met. **87**, 53 (1982)
7. S. Mühlbauer, B. Binz, F. Jonietz, C. Pfleiderer, A. Rosch, A. Neubauer, R. Georgii, P. Böni, Science **323**, 915 (2009)
8. M. Uchida, Y. Onose, Y. Matsui, Y. Tokura, Science **311**, 359 (2006)
9. P. Bak, M.H. Jensen, J. Phys. C **13**, L881 (1980)
10. O. Nakanishi, A. Yanase, A. Hasegawa, M. Kataoka, Solid State Commun. **35**, 995 (1980)
11. M. Kataoka, J. Phys. Soc. Jpn. **56**, 3635 (1986)
12. S.V. Grigoriev, S.V. Maleyev, A.I. Okorokov, Y.O. Chetverikov, H. Eckerlebe, Phys. Rev. B **73**, 224440 (2006)
13. S.V. Grigoriev, V.A. Dyadkin, D. Menzel, J. Schoenes, Y.O. Chetverikov, A.I. Okorokov, H. Eckerlebe, S.V. Maleyev, Phys. Rev. B **76**, 224424 (2007)
14. M. Takeda, Y. Endoh, K. Kakurai, Y. Onose, J. Suzuki, Y. Tokura, J. Phys. Soc. Jpn. **78**, 093704 (2009)
15. K. Ishizuka, B. Allman, J. Electron Microsc. **54**, 191 (2005)
16. J. Beille, J. Voiron, M. Roth, Solid State Commun. **47**, 399 (1983)
17. N. Kanazawa, Y. Onose, T. Arima, D. Okuyama, K. Ohoyama, S. Wakimoto, K. Kakurai, S. Ishiwata, Y. Tokura, Phys. Rev. Lett. **106**, 156603 (2011)
18. N. Kanazawa, J.-H. Kim, D.S. Inosov, J.S. White, N. Egetenmeyer, J.L. Gavilano, S. Ishiwata, Y. Onose, T. Arima, B. Keimer, Y. Tokura, Phys. Rev. B **86**, 134425 (2012)
19. Y. Li, N. Kanazawa, X.Z. Yu, A. Tsukazaki, M. Kawasaki, M. Ichikawa, X. F. Jin, F. Kagawa and Y. Tokura. Phys. Rev. Lett. **110**, 117202 (2013)
20. S.V. Grigoriev, V.A. Dyadkin, E.V. Moskvin, D. Lamago, Th Wolf, H. Eckerlebe, S.V. Maleyev, Phys. Rev. B **79**, 144417 (2009)
21. Y. Onose, N. Takeshita, C. Terakura, H. Takagi, Y. Tokura, Phys. Rev. B **72**, 224431 (2005)
22. K. Shibata, X.Z. Yu, T. Hara, D. Morikawa, N. Kanazawa, K. Kimoto, S. Ishiwata, Y. Matsui, Y. Tokura, Nat. Nanotechnol. **8**, 723 (2013)
23. B. Lebech, J. Bernhard, T. Freltoft, J. Phys.: Condens. Matter **1**, 6105 (1989)
24. T. Adams, A. Chacon, M. Wagner, A. Bauer, G. Brandl, B. Pedersen, H. Berger, P. Lemmens, C. Pfleiderer, Phys. Rev. Lett. **108**, 237204 (2012)
25. S. Seki, X.Z. Yu, S. Ishiwata, Y. Tokura, Science **336**, 198 (2012)
26. S. Kusaka, K. Yamamoto, T. Komatsubara, Y. Ishikawa, Solid State Commun. **20**, 925 (1976)
27. M. Date, K. Okuda, K. Kadowaki, J. Phys. Soc. Jpn. **42**, 1555 (1977)
28. K. Kadowaki, K. Okuda, M. Date, J. Phys. Soc. Jpn. **51**, 2433 (1981)
29. A. Bogdanov, D.A. Yablonskii, Sov. Phys. JETP **68**, 101 (1989)
30. A. Bogdanov, A. Hubert, J. Magn. Magn. Mater. **138**, 255 (1994)
31. X.Z. Yu, Y. Onose, N. Kanazawa, J.H. Park, J.H. Han, Y. Matsui, N. Nagaosa, Y. Tokura, Nature (London) **465**, 901 (2010)
32. T.H.R. Skyrme, Proc. R. Soc. A **260**, 127 (1961); T.H.R. Skyrme. Nucl. Phys. **31**, 556 (1962)
33. C. Pfleiderer, D. Reznik, L. Pintschovius, H.v. Löhneysen, M. Garst, A. Rosch. Nature **427**, 227 (2004)
34. U.K. Rößler, A.N. Bogdanov, C. Pfleiderer, Nature **442**, 797 (2006)
35. B. Binz, A. Vishwanath, V. Aji, Phys. Rev. Lett. **96**, 207202 (2006)
36. I. Fischer, N. Shah, A. Rosch, Phys. Rev. B **77**, 024415 (2008)
37. O. Nakanishi, A. Yanase, A. Hasegawa, J. Magn. Magn. Mater. **15–18**, 879 (1980)
38. L.F. Mattheiss, D.R. Hamann, Phys. Rev. B **47**, 13114 (1993)
39. A. Sakai, F. Ishii, Y. Onose, Y. Tomioka, S. Yotsuhashi, H. Adachi, N. Nagaosa, Y. Tokura, J. Phys. Soc. Jpn. **76**, 093601 (2007)

40. S. Yeo, S. Nakatsuji, A.D. Bianchi, P. Schlottmann, Z. Fisk, L. Balicas, P.A. Stampe, R.J. Kennedy, Phys. Rev. Lett. **91**, 046401 (2003)
41. V.I. Anisimov, R. Hlubina, M.A. Korotin, V.V. Mazurenko, T.M. Rice, A.O. Shorikov, M. Sigrist, Phys. Rev. Lett. **89**, 257203 (2002)
42. A.K. Geim, K.S. Novoselov, Nat. Mater. **6**, 183 (2007)
43. M.Z. Hasan, C.L. Kane, Rev. Mod. Phys. **82**, 3045 (2010)

Chapter 2
Experimental Methods

2.1 Bulk Crystal Growth

2.1.1 Polycrystal Samples

Polycrystalline samples of B20-type germanides were synthesized with a cubic-anvil-type high-pressure apparatus (see also Ref. [1]). A mixture of elemental materials with an atomic ratio of T : Ge $= 1 : 1$ was at first arc-melted in an argon atmosphere. (Here, T represents transition-metal element.) The alloy was placed in a cylindrical BN capsule (Fig. 2.1) and was heat-treated for 1 hour at 1073 K under a high pressure of 4.0 GPa. Some samples include a trace of impurity less than 4 % in volume fraction.

Figure 2.2 shows some powder X-ray diffraction (XRD) patterns and composition dependence of lattice constant of synthesized B20-type germanides. Monotonous change in the lattice constant indicates good solubility of the transition-metal elements.

2.1.2 FeGe Single Crystal

A single crystal of FeGe was grown by chemical vapor transport (CVT) method. A mixture of Fe and Ge powders with an atomic ratio of 1 : 1 was put into a cylindrical quartz tube, which was evacuated below 1×10^{-4} Torr. The evacuated tube was heated at 700 °C for 1 day. The obtained powder mainly consists of FeGe with B35 crystal structure. The powder of B35-type FeGe was placed with I_2 (20 mg) in an evacuated cylindrical quartz tube whose volume is $\pi 10^2 \times 120$ mm^3. The sample was placed in a three-zone furnace and heated for 1 month with a thermal gradient: 560 °C at an end of cylinder where powder sample was placed and 500 °C at the other end of the cylinder. Many pieces of B20-type FeGe single crystals were grown

N. Kanazawa, *Charge and Heat Transport Phenomena in Electronic and Spin Structures in B20-type Compounds*, Springer Theses,
DOI 10.1007/978-4-431-55660-2_2

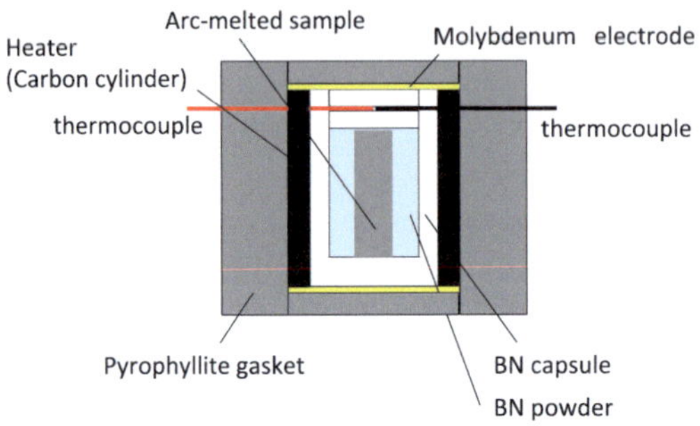

Fig. 2.1 Cross-section drawing of assembly for high-pressure synthesis. A high pressure is applied in six directions of the cubic gasket surface

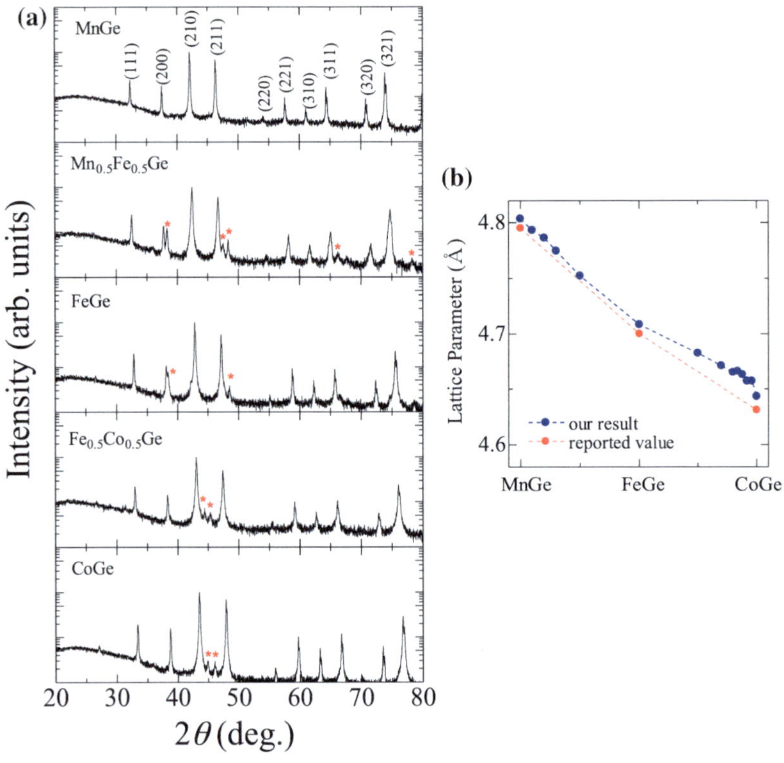

Fig. 2.2 **a** Several X-ray diffraction patterns of the synthesized B20-type germanides. *Red* asterisks indicate intensities from impurity. **b** Calculated lattice constants from the XRD patterns in comparison with previous reports [1]. A linear relation between the composition ratio and the lattice constant verifies good solubility of Mn, Fe, and Co

at the lower temperature side of the cylinder. Typical size of the crystals is 0.7 mm. The crystals have cleavage planes. Figure 2.3 shows the single crystal with (111) cleavage plane, which was used for measurements.

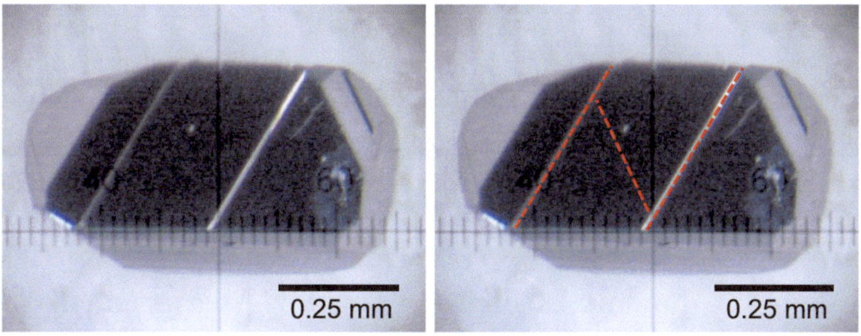

Fig. 2.3 B20-type FeGe single crystal grown by CVT method. The wide surface is a cleavage plane of (111) surface with small steps (*red dashed lines*) forming regular triangles

2.2 Thin-Film Growth

FeGe epitaxial films with 40-nm thickness were grown on highly-resistive Si(111) substrates (ρ > 1000 Ω cm) by molecular beam epitaxy. Figure 2.4 shows lattice matching between FeGe(111) surface and Si(111) surface. Fe(Ge) atoms on the green and purple triangles indicated in Fig. 2.4b, d are stacked on corresponding Si atoms indicated in Fig. 2.4c. The lattice mismatch is −0.05 %.

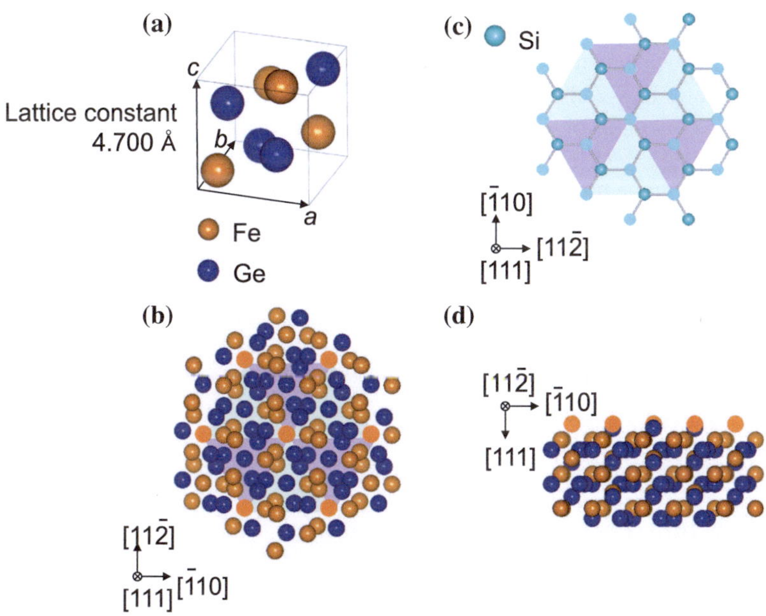

Fig. 2.4 Schematics of atomic arrangements of **a** a unit cell of FeGe, **b** FeGe viewed from [111] crystal axis, **c** Si viewed from [111] crystal axis, and **d** FeGe viewed from [11$\bar{2}$] crystal axis

Fe and Ge were deposited on the wafer using Knudsen cells. We deposited Fe and Ge onto a Si(111)-7 × 7 surface heated at 325 °C by using Knudsen cells until a prescribed thickness was reached. The characteristic $\sqrt{3} \times \sqrt{3}$ reflection high-energy electron diffraction (RHEED) pattern on the B20-compound (111) surface appeared during the deposition. The single phase nature was confirmed by 2θ–θ XRD.

2.3 Magnetization and Electrical Transport Measurement

Samples for magnetization and electrical transport measurements were cut into rectangular shape with a typical size of $4 \times 1.5 \times 0.2\ \text{mm}^3$. Magnetization and electrical transport measurements were performed using Quantum Design Physical Property Measurement System (PPMS) and Magnetic Property Measurement System (MPMS). Resistivity was measured with a four-probe method. Measurements of magnetoresistance and Hall resistivity were performed with the current parallel to the longest side (x axis) and the magnetic field to the shortest one (z axis). Electrical contacts were made with indium. Magnetization (M) was measured with the same magnetic field direction as the transport measurements.

2.4 Neutron Scattering

High-angle powder neutron diffraction experiments on MnGe were performed using HERMES and TAS-2 installed at the JRR-3 reactor of Japan Atomic Energy Agency (JAEA). Wavelengths of used neutron beams at HERMES and TAS-2 are 1.8449(1) Å and 2.359 Å, respectively. The powder sample includes a trace of impurity less than 4 % in volume fraction.

Small-angle neutron scattering experiments on MnGe were carried out using SANS-I instrument at the Paul Scherrer Institut using neutrons with wavelength 4.7 Å. The same powder sample was packed in a single-crystalline silicon container with a height of 9.5 mm and a diameter of 18 mm, and installed in a cryomagnet. The magnetic field was applied perpendicular to the incident neutron beam collimated over 4.5 m. The diffracted neutrons were collected by a two-dimensional multidetector placed 2 m behind the sample.

2.5 Thermal Transport Measurement

Seebeck coefficients and thermal conductivities were measured in a setup placed in a vacuum ($<1 \times 10^{-4}$ Torr) (Fig. 2.5). One end of the sample was fixed to a copper block as the heat bath by silver epoxy, and the other to an 10 kΩ resistor through which electric current was passed to generate heat current. Chromel–constantan

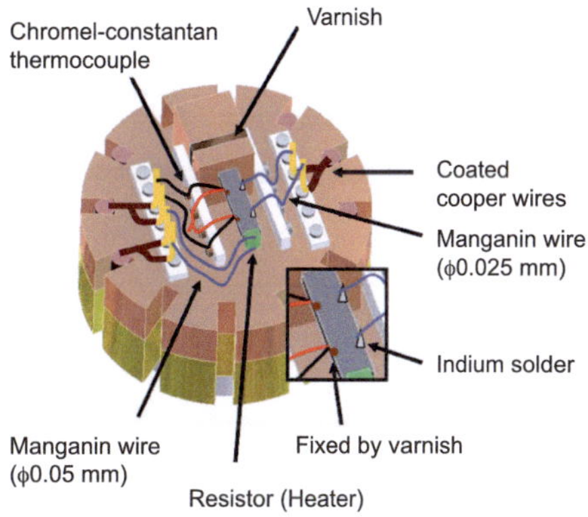

Fig. 2.5 Setup for thermal transport measurement

thermocouples were attached to a sample by varnish. The thermal electromotive force was measured using manganin wires, whose Seebeck coefficient is negligible small (less than 1 μV/K below 300 K) [2].

References

1. H. Takizawa, T. Sato, T. Endo, M. Shimada, J. Solid State Chem. **73**, 40 (1988)
2. K.D.D. Rathnayaka, J. Phys. E: Sci. Instrum. **18**, 380 (1985)

Chapter 3
Magnetic and Transport Properties in B20-type Germanides

3.1 Introduction

The Hall effect, a transverse electromagnetic response to an applied longitudinal stimulation, is the main issue in this and the following two chapters. In particular, we focus on roles of "emergent electromagnetic fields" inherent in topological magnetic arrangements [1] in the Hall effects. Unique Hall responses can be observed according to each topological quantity in each magnetic structure, which conversely means the Hall effect is a suitable probe for magnetic structures. In this section, we introduce a brief review on the Hall effect induced by the emergent electromagnetic field.[1]

3.1.1 Berry Phase and Emergent Electromagnetic Field

Edwin Herbert Hall discovered the Hall effect arising from the spontaneous magnetization in metallic ferromagnetic systems, which is called the anomalous Hall effect (AHE) [2]. The mechanism has been a controversial issue for more than a century, though it is commonly recognized that the spin-orbit interaction is the origin of AHE [3]. A recent great progress in the issue is the Berry-phase scenario, which can be applicable to many types of Hall effects and furthermore to various phenomena in the quantum systems [3, 4]. Here we introduce the Berry phase [5] as the source of a emergent electromagnetic field [4].

Here, we imagine an infinitesimal translation of a vector bound on a sphere. This infinitesimal translation means the projection of the vector onto a tangent plane after a translation in the 3D free space. The direction of the vector changes after a sequential

[1] A comprehensive and detail review of physical phenomena associated with emergent electromagnetic fields is provided by Ref. [1].

N. Kanazawa, *Charge and Heat Transport Phenomena in Electronic and Spin Structures in B20-type Compounds*, Springer Theses,
DOI 10.1007/978-4-431-55660-2_3

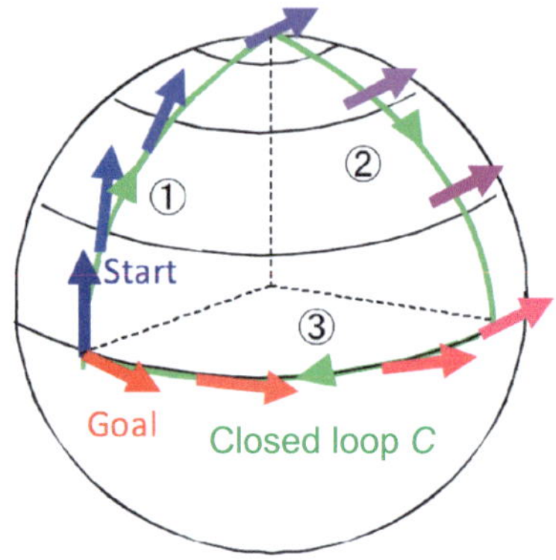

Fig. 3.1 Conceptual diagram of the Berry phase

translation along a closed loop C as shown in Fig. 3.1. This is due to the constraint, i.e., sticking the vector on the "curved" surface. The variation of the vector direction is written by the integral of the curvature in the area enclosed by the loop C, which is called as holonomy in the topology.

The same thing happens to the state vector in quantum mechanics, i.e., the state vector, the subspace of the Hilbert space, and the Hilbert space respectively correspond to the vector, the sphere, and the 3D free space appearing in the above example. As the direction of the state vector represents the quantum phase, an additional phase factor, corresponding to the variation of the vector direction in the above example, can be multiplied to the state via an adiabatic parameter change. This additional phase is so-called Berry phase.

More precisely, the Berry phase is defined as follows [4]. The Schrödinger equation in the system where the parameter set $\mathbf{R}(t)$ varies with time is

$$H(\mathbf{R}(t))|n(\mathbf{R}(t))\rangle = E_n(\mathbf{R}(t))|n(\mathbf{R}(t))\rangle. \tag{3.1}$$

$|n(\mathbf{R}(t))\rangle$ is the eigenstate of energy. The state vector of the electron at time t ($|\phi_n(t)\rangle$) is assumed to be

$$|\phi_n(t)\rangle = \exp(i\gamma_n(t)) \exp\left[-\frac{i}{\hbar}\int_0^t dt' E_n(\mathbf{R}(t'))\right] |n(\mathbf{R}(t))\rangle, \tag{3.2}$$

where two phase factors are multiplied; the first exponential results in the Berry phase and the second represents the time evolution. This state vector $|\phi_n(t)\rangle$ should satisfy the time-dependent Schrödinger equation $i\hbar\frac{\partial}{\partial t}|\phi_n(t)\rangle = H(\mathbf{R}(t))|\phi_n(t)\rangle$. The nontrivial phase factor γ_n is derived under this constraint as

$$\gamma_n = \int_C d\mathbf{R} \cdot \mathscr{A}_n, \tag{3.3}$$

$$\mathscr{A}_n(\mathbf{R}) = i\langle n(\mathbf{R})|\frac{\partial}{\partial \mathbf{R}}|n(\mathbf{R})\rangle. \tag{3.4}$$

$\mathscr{A}_n$ is referred to as the Berry connection or the Berry vector potential. If the path C which $\mathbf{R}(t)$ tracks is not closed, a certain gauge transformation makes γ_n to be zero. If the path C is closed, γ_n is single-valued and gauge invariant (Fig. 3.2).

Furthermore, a gauge invariant "Berry curvature $\Omega_n(\mathbf{R})$" is defined as[2]

$$\Omega_n(\mathbf{R}) = \nabla_{\mathbf{R}} \times \mathscr{A}_n(\mathbf{R}), \tag{3.5}$$

$$\gamma_n = \int_S d\mathbf{S} \cdot \Omega_n(\mathbf{R}). \tag{3.6}$$

The Berry curvature can be regarded as the magnetic field emerging in the intended parameter-space because the mathematical construction is same with the classical magnetic field.

3.1.2 Correction of Semi-classical Transport Equations—Berry Phase in the Momentum Space

The semi-classical transport equations need to be corrected in consideration of the Berry phase. Here we only consider the Berry phase emerging in the momentum space (a band structure). The group velocity in the electron system under an application of an electric field $\mathbf{E}$ is written by

$$\mathbf{v}_n(\mathbf{k}) = \frac{\partial E_n(\mathbf{k})}{\hbar \partial \mathbf{k}} - \frac{e}{\hbar}\mathbf{E} \times \Omega_n(\mathbf{k}), \tag{3.7}$$

$$\Omega_n(\mathbf{k}) = i\langle \nabla_{\mathbf{k}} u_n(\mathbf{k})| \times |\nabla_{\mathbf{k}} u_n(\mathbf{k})\rangle. \tag{3.8}$$

Here the gauge-invariant crystal-momentum is $\mathbf{k} = \mathbf{q} + \frac{e}{\hbar}\mathbf{A}(t)$ and the vector potential of the classical electromagnetic field is $\mathbf{A}(t)$. The term including the Berry curvature in Eq. 3.7 is called an "anomalous velocity", which induces a Hall current perpendicular to the applied electric field $\mathbf{E}$. In magnets, the spin-orbit interaction associated with the spontaneous magnetization invokes a band splitting, which results in the finite Berry phase. This is the Berry-phase scenario of AHE.

Eventually, the semi-classical transport equations in the presence of the Berry phase in the momentum space are

$$\dot{\mathbf{r}} = \frac{\partial E_n(\mathbf{k})}{\hbar \partial \mathbf{k}} - \dot{\mathbf{k}} \times \Omega(\mathbf{k}), \tag{3.9}$$

$$\hbar \dot{\mathbf{k}} = -e\mathbf{E} - e\dot{\mathbf{r}} \times \mathbf{B}. \tag{3.10}$$

[2]Here we assume a 3D parameter space for simplicity.

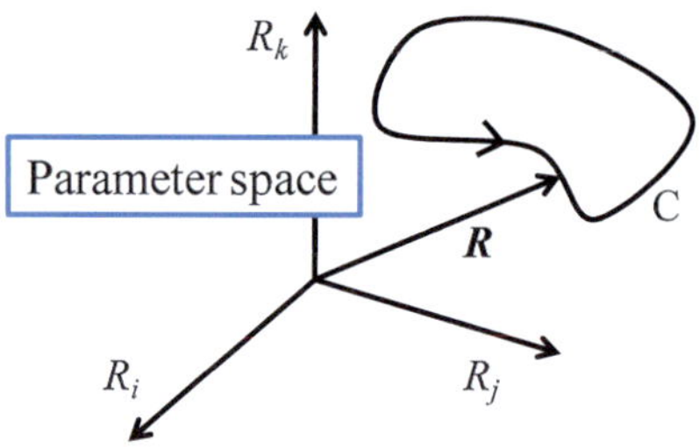

Fig. 3.2 An adiabatic change in parameters which the Hamiltonian depends on

In addition, the derived formula of the Hall conductivity is described as

$$\sigma_{xy} = \frac{e^2}{\hbar}\int \frac{d\mathbf{k}}{(2\pi)^3} f(E_{\mathbf{k}})\Omega^z. \tag{3.11}$$

The important point is that the anomalous velocity, and equally the Hall conductivity, are independent of the relaxation time τ.

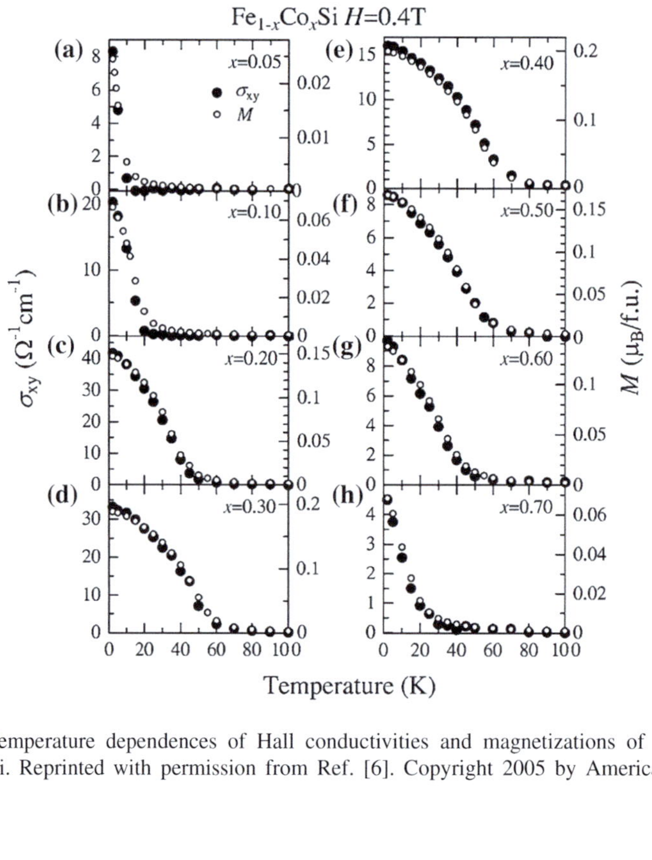

Fig. 3.3 Temperature dependences of Hall conductivities and magnetizations of a series of $Fe_{1-x}Co_xSi$. Reprinted with permission from Ref. [6]. Copyright 2005 by American Physical Society

3.1.3 Anomalous Hall Effect in B20-type Silicides

A comprehensive study of anomalous Hall effect in a series of B20-type $Fe_{1-x}Co_xSi$ was reported by Onose et al. [6], and the Berry-phase scenario is definitely verified in MnSi by Lee et al. [7]. Figure 3.3 shows the T-dependence of Hall conductivity σ_{xy} and magnetization M of $Fe_{1-x}Co_xSi$ at $H = 0.4$ T. Both the two T-dependences (σ_{xy}-T and M-T) trace each other, which indicates the relation $\sigma_{xy} = S_H M$, i.e., the Hall current is independent of τ as demonstrated by the Berry-phase scenario.

The τ-independent Hall current is presented in exact detail by examining various σ_{xy}-H curves in MnSi. Lee et al. employed the following relation [7]:

$$\sigma_{xy} = \sigma_{xy}^{\mathrm{N}} + \sigma_{xy}^{\mathrm{A}} = R_0 \rho_{xx}^{-2} H + S_{\mathrm{H}} M, \tag{3.12}$$

which can be converted to

$$\rho_{yx} = \rho_{yx}^{\mathrm{N}} + \rho_{yx}^{\mathrm{A}} = R_0 H + S_{\mathrm{H}} \rho_{xx}^2 M \tag{3.13}$$

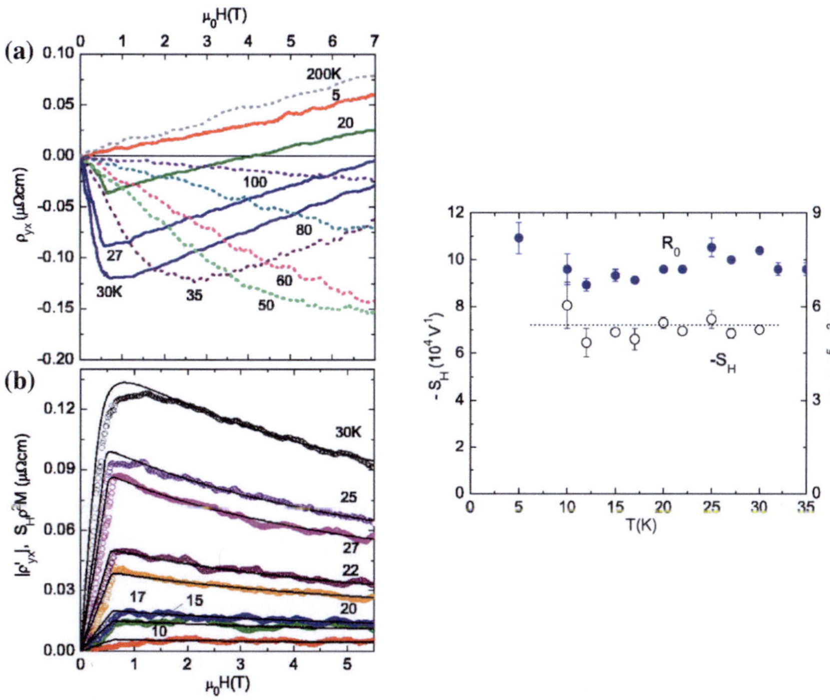

Fig. 3.4 (*left-top*) Magnetic-field dependence of Hall resistivity in MnSi, (*left-bottom*) comparison of the extracted anomalous Hall component and the fitting curve (see also text), and (*right*) the fitting parameters of normal Hall coefficient R_0 and anomalous Hall coefficient S_H, which are both insensitive to temperature. Reprinted with permission from Ref. [7]. Copyright 2007 by American Physical Society

if the Hall angle is sufficiently small ($\rho_{yx} \ll \rho_{xx}$). The measured Hall resistivity is well reproduced by the Eq. 3.13 with use of the two-parameter fitting (R_0 and S_H). The constancy of S_H against a large change (one order of magnitude) in the longitudinal conductivity $\sigma_{xx} (\propto \tau)$ fully supports the Berry-phase scenario (Fig. 3.4).

3.2 Magnetization Properties

We can grab a rough outline of the magnetic ordering in B20-type germanides from magnetization measurements. Typical magnetization profiles appearing in B20-type silicides are observed in $Mn_{1-x}Fe_xGe$ ($0.4 \leq x \leq 1$) and $Fe_{1-y}Co_yGe$ ($0 \leq y \leq 0.84$), while much different ones appear in $Mn_{1-x}Fe_xGe$ ($0 \leq x \leq 0.3$). Distinct magnetization properties in $Mn_{1-x}Fe_xGe$ ($0 \leq x \leq 0.3$) probably originate from much enlarged DM interactions, which also lead to short-period helical structures (several nanometers).

Figure 3.5 shows temperature dependence of magnetization at $H = 100$ Oe in $Mn_{1-x}Fe_xGe$ and $Fe_{1-y}Co_yGe$. The M-T curves are categorized in three types: (i) a curve with a broad antiferromagnetic peak [$Mn_{1-x}Fe_xGe$ ($0 \leq x \leq 0.3$)], (ii) a stepwise curve with two sharp inflections, which represent the helical ordering (the lower temperature) and the onset of the development of magnetic correlation (the higher temperature) [$Mn_{1-x}Fe_xGe$ ($0.4 \leq x \leq 1$) and $Fe_{1-y}Co_yGe$ ($0 \leq y \leq 0.84$)]. (iii) a typical paramagnetic M-T curve (CoGe). Type-(ii) M-T curves are similar with those of the B20-type silicide magnets.

The transition temperature, more precisely the helical-ordering temperature, T_N is defined as (i) the temperature at which the magnetization starts to decrease in decreasing temperature process [$Mn_{1-x}Fe_xGe$ ($0 \leq x \leq 0.3$)] or (ii) the lower temperature at which the stepwise M-T curve bends [$Mn_{1-x}Fe_xGe$ ($0.4 \leq x \leq 1$) and $Fe_{1-y}Co_yGe$ ($0 \leq y \leq 0.84$)].

Figure 3.6 presents magnetic field dependences of magnetizations at various temperatures in $Mn_{1-x}Fe_xGe$. In $Mn_{1-x}Fe_xGe$ ($0 \leq x \leq 0.3$), the slope of M-H curve below H_c becomes more moderate with decreasing temperature, while those in $Mn_{1-x}Fe_xGe$ ($0.4 \leq x \leq 1$) remain constant. The theoretical study introduced in Sect. 1.2 predicts the latter magnetization profile (the constant slope) [8].

Another distinct magnetic property in $Mn_{1-x}Fe_xGe$ ($0 \leq x \leq 0.3$) also appears in the magnitude of H_c. The critical field H_c at $T = 5$ K of $Mn_{1-x}Fe_xGe$ ($0 \leq x \leq 0.3$) is one-order larger than those of $Mn_{1-x}Fe_xGe$ ($0.4 \leq x \leq 1$) and other B20-type magnets. The enhanced H_c represents quite a large DM interaction.

A collaborative study on real-space observation by LTEM found formations of helical and skyrmion structures in $Mn_{1-x}Fe_xGe$ ($0.4 \leq x \leq 1$) [9]. Figure 3.7 shows representative LTEM images of helical and skyrmion structures observed in some B20-type germanides. The helical period λ and the skyrmion size a_{sk} are estimated from the period of the LTEM contrast patterns. In addition, the correlation between LTEM patterns and convergent-beam-electron-diffraction (CBED) patterns determines the helicity of magnetic structures (spin rotation direction in helical structure

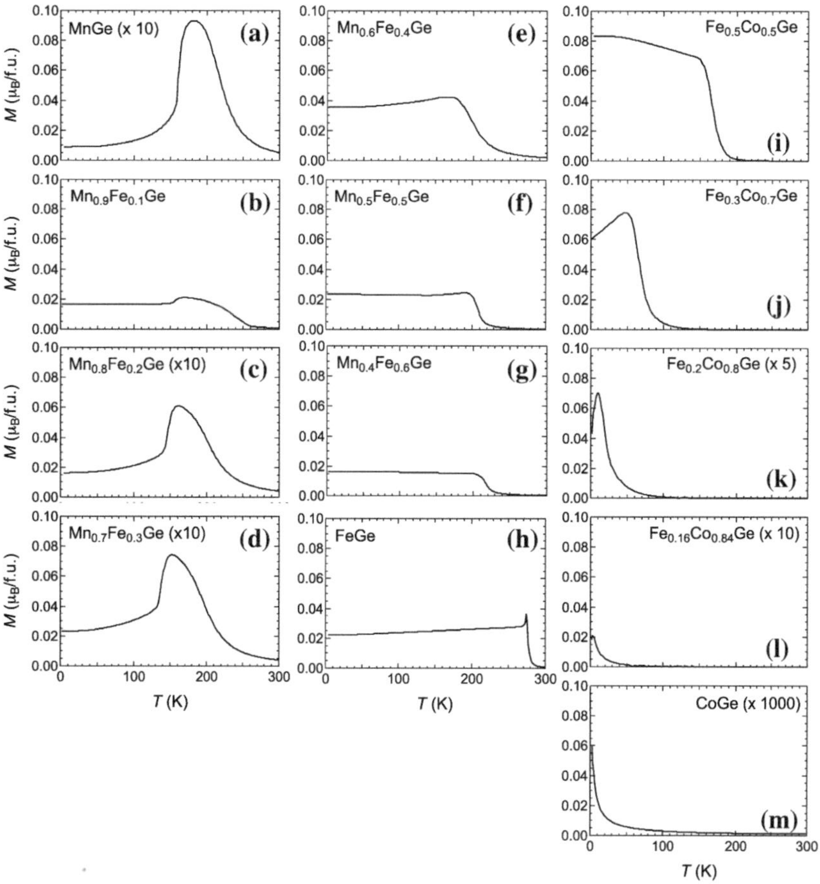

Fig. 3.5 Temperature dependences of magnetizations in a series of B20-type germanides at $H = 100$ Oe

and spin swirling direction in skyrmion structure), which corresponds to the sign of DM interaction.[3] The right-bottom panel of Fig. 3.7 shows the composition dependence of the helical period λ or the corresponding skyrmion crystal lattice constant a_{sk} multiplied by $\sqrt{3}/2$ with the sign of the DM interaction (denoted by $\Gamma_c \times \gamma_m$). A diverging behavior of the helical period and a sign reversal of the DM interaction are observed in the vicinity of $x = 0.8$, which we attribute to the zero crossover point in the DM interaction: The right-top panel in Fig. 3.7 shows a rough estimation of the strength of the DM interaction or spin-orbit interaction. From Eqs. 1.1 and 1.2, we can evaluate the strength of the DM interaction $|D|$ as T_{N}/λ. Again the diverging behavior of $\lambda = J/|D|$ is clearly understood.

---

[3]For details, see Ref. [9].

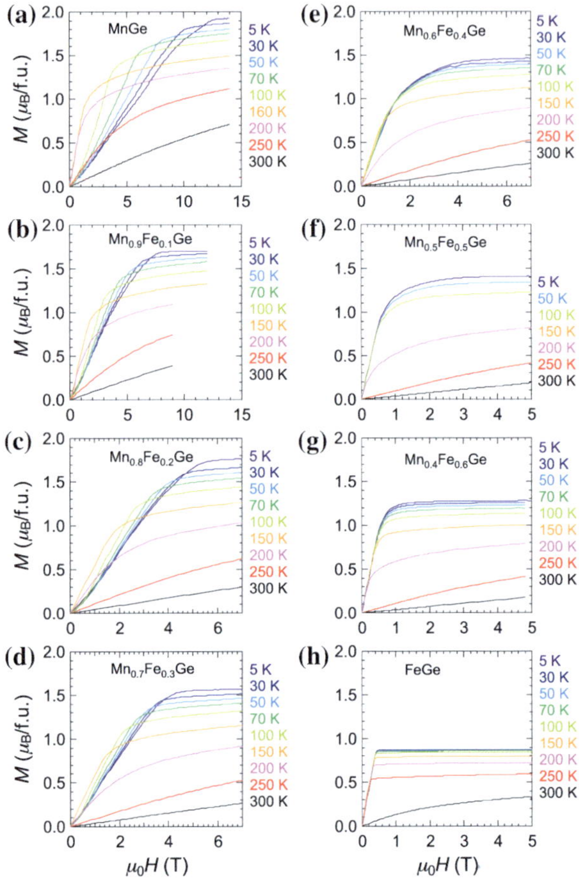

Fig. 3.6 Magnetic-field dependences of magnetizations in B20-type $Mn_{1-x}Fe_xGe$ at various temperatures

We summarize the magnetization properties of B20-type germanides in Fig. 3.8 with reference data of B20-type silicides [6, 10, 11]. There seems to exist a magnetic phase boundary around $Mn_{1-x}Fe_xGe$ ($x = 0.3$–0.4): Variations of T_N and H_c with the transition-metal composition indicate kinks (a dip and a sharp reduction, respectively) around the corresponding composition (Fig. 3.8a, b). This phase separation may originate from a large change in the DM interaction, whose influence can be seen in the huge variation of λ: 3 nm for MnGe [12, 13] and 220 nm for $Mn_{0.15}Fe_{0.85}Ge$ [9].

Here we employ a toy model to roughly estimate the magnitude of the DM interaction. The energy of the helical structure is $E = \sum(J\mathbf{S}_i \cdot \mathbf{S}_j + D\mathbf{S}_i \times \mathbf{S}_j) = \sum(JS^2 \cos\theta + DS^2 \sin\theta)$. The state producing the minimum energy is realized: $\partial E/\partial\theta = S^2(-J\sin\theta + D\cos\theta) = 0$ holds, and therefore $J/D = \cos\theta/\sin\theta$. In the case of MnGe with $\lambda = 3$ nm, the angle between two spins which are positioned at one-unit-cell length ($\approx$5 Å) from each other is approximately 60°, i.e., $D \approx J\sin(60°)/\cos(60°) = \sqrt{3}J$. This result violates the assumption of the theo-

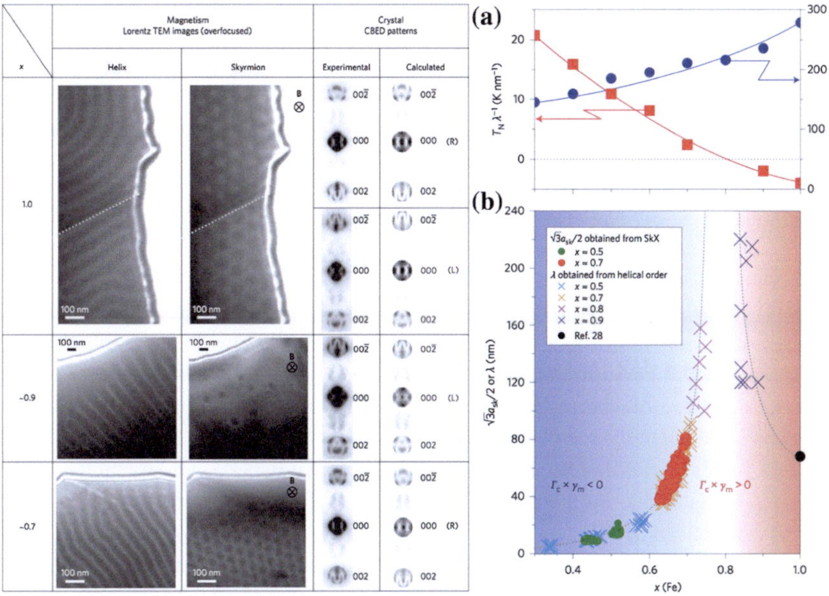

Fig. 3.7 (*left*) Typical LTEM images of helical and skyrmion structures observed in some B20-type germanides with their convergent-beam electron diffraction patterns. Iron content x dependence of (*right-top*) T_N/λ (a barometer of the DM interaction D), T_N (a barometer of exchange interaction J), and (*right-bottom*) the helical period ($\lambda = \sqrt{3}a_{sk}/2$). Reprinted by permission from Macmillan Publishers Ltd: Ref. [9], copyright 2013

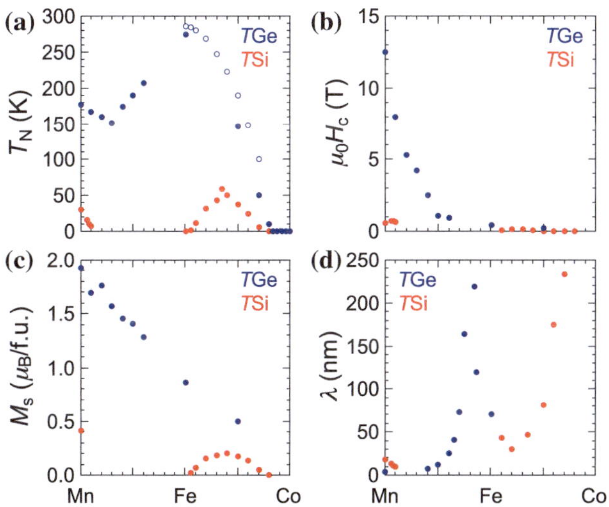

Fig. 3.8 Variations of **a** the helical ordering temperature T_N, **b** the critical field H_c, **c** the saturated magnetization M_s, and **d** the helical period of B20-type germanides as a function of the transition-metal element. The adscript data of B20-type silicides (*red solid circles*) and B20-type $Fe_{1-x}Co_xGe$ (*blue open circles*) are reproduced from Refs. [6, 10, 11]

retical model for B20-silicide magnets [14–16], i.e., $D \ll J$. The perturbative expansion in $Q = D/J$ is not allowed in the case of MnGe. The large DM interaction may trigger a different magnetic texture in short-period helimagnets $Mn_{1-x}Fe_xGe$ $(0 \leq x \leq 0.3)$.

3.3 Transport Properties

We mainly focus on the anomalous Hall effect in $Mn_{1-x}Fe_xGe$ in this section. The composition x dependence of the anomalous Hall coefficient S_H resembles that of the strength of the DM interaction defined as $T_N/(\lambda M_s^2)$; the clear proportionality between the spin-orbit interaction and the anomalous Hall current is demonstrated, which has been scarcely confirmed experimentally.

Figure 3.9 shows temperature dependences of resistivities ρ_{xx} and mean free paths l in $Mn_{1-x}Fe_xGe$. The mother compounds MnGe and FeGe demonstrate large residual resistivity ratios (RRR) of 19 and 35, respectively. $Mn_{0.8}Fe_{0.2}Ge$ shows more than twice higher resistivity compared to the other compounds, which indicates bad sample quality. The mean free path is calculated by $\sigma_{xx} = \frac{e^2 k_F^2}{3\pi^2 \hbar} l$, where the Fermi wave vector $k_F = (3\pi^2 n)^{1/3}$ and the carrier density n is set to be $1.25 \times 10^{22}\,\text{cm}^{-3}$. While MnGe and FeGe demonstrate a relatively long l comparable to their helical periods of 3 and 70 nm, the other compounds shows almost equal to or less than 1 nm.

Figure 3.10 shows magnetic-field dependences of magnetoresistivities $\rho_{xx}(H)/\rho_{xx}(0)$ in $Mn_{1-x}Fe_xGe$. The profile of magnetoresistivity drastically changes across the possible phase boundary at $x = 0.3$–0.4. The $\rho_{xx}(H)/\rho_{xx}(0)$ in $Mn_{1-x}Fe_xGe$ $(0 \leq x \leq 0.2)$ changes by some 10 % when ramping the magnetic field to H_c, whereas the change by the order of 1 % is observed in $Mn_{1-x}Fe_xGe$ $(0.4 \leq x \leq 1)$.

We present magnetic-field dependences of Hall resistivities ρ_{yx} for the two magnetic phases separated by $x \approx 0.3$ in Figs. 3.11 and 3.12, respectively. By using

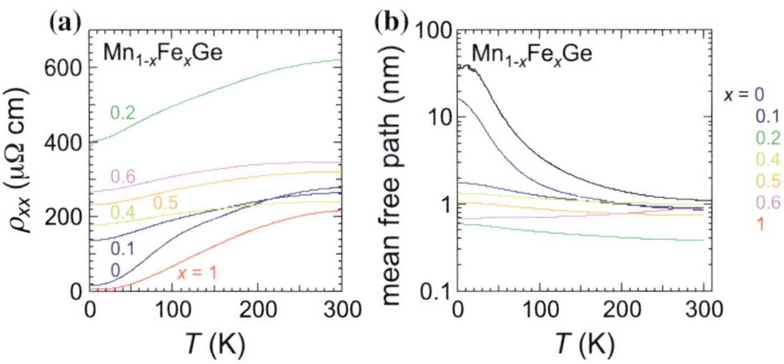

Fig. 3.9 Temperature dependence of **a** resistivity and **b** mean free path at zero magnetic field in $Mn_{1-x}Fe_xGe$. The mean free path l is estimated from the relation $\sigma_{xx} = \frac{e^2 k_F^2}{3\pi^2 \hbar} l$ (see also text)

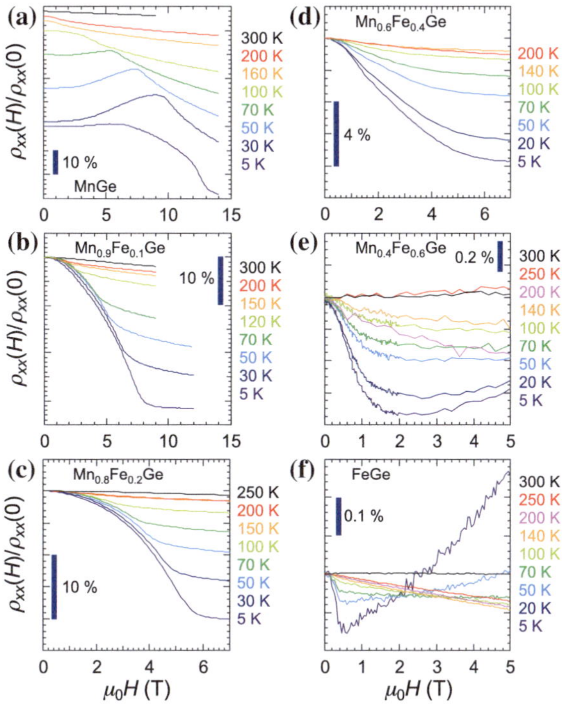

Fig. 3.10 Magnetic-field dependences of magnetoresistivities at various temperatures in $Mn_{1-x}Fe_xGe$

the data of M and $\rho_{xx}(H)$, we can estimate the normal and the anomalous Hall coefficients R_0 and S_H. Here we note that we have to take the large variations of $\rho_{xx}(H)/\rho_{xx}(0)$ into account when analyzing the AHE because the anomalous Hall resistivity includes the term proportional to $\rho_{xx}^2 M$ (Eq. 3.13). As discussed in Sect. 3.1, Lee et al. successfully explained the AHE in MnSi by employing the relation Eq. 3.13, which is empirically validated for the ferromagnetic (spin-collinear) system [7]. Here, we fit the measured ρ_{yx} above H_c, where the spin-collinear state is realized, by evaluating R_0 and S_H. If there is a difference between the measured ρ_{yx}-H curve and the empirically anticipated curve in the helical phase below H_c, the difference will be an additional contribution to Hall effect originating from a formation of a non-coplanar magnetic structure.

We can determine the Hall coefficients R_0 and S_H as the intercept and the slope of the curve ρ_{yx}/H versus $\rho_{xx}^2 M/H$ above the critical field H_c.[4] The fitting curves are also indicated with the measured Hall resistivities in Figs. 3.11 and 3.12. The discrepancy between them can be clearly observed in ρ_{yx} at low temperatures in MnGe. In the other compounds, the AHE relation Eq. 3.13 well reproduces ρ_{yx} at all the temperatures below T_N. As the helical period λ is almost constant in the composition range of $0.1 \leq x \leq 0.4$ (Fig. 3.8), the discrepancy observed in MnGe

[4]The Eq. 3.13 can be converted to $\frac{\rho_{yx}}{H} = R_0 + S_H \frac{\rho_{xx}^2 M}{H}$ by divided by H.

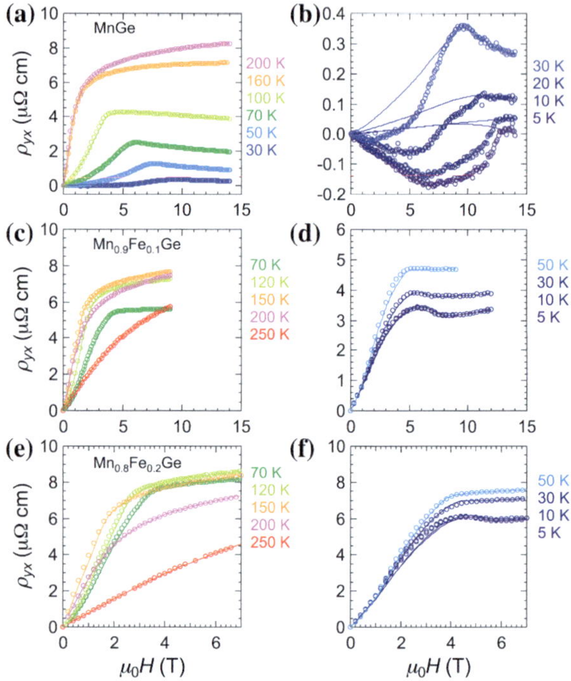

Fig. 3.11 Magnetic-field dependences of Hall resistivities in $Mn_{1-x}Fe_xGe$ ($0 \leq x \leq 0.2$). *Solid lines* in all the panels are the fitted curves of ρ_{yx} using the relation that $\rho_{yx} = R_0 H + S_H \rho_{xx}^2 M$ with the fitting parameters R_0 and S_H (see text). *Left* and *right* panels contain data at high and low temperatures, respectively

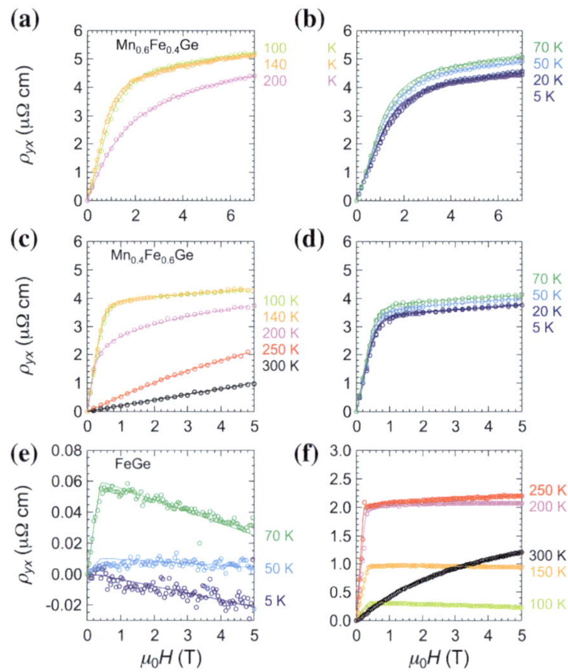

Fig. 3.12 Magnetic-field dependences of Hall resistivities in $Mn_{1-x}Fe_xGe$ ($0.4 \leq x \leq 1$). *Solid lines* in all the panels are the fitting curves of ρ_{yx} using the relation that $\rho_{yx} = R_0 H + S_H \rho_{xx}^2 M$ with the fitting parameters R_0 and S_H (see text). *Left* and *right* panels contain data at high and low temperatures, respectively

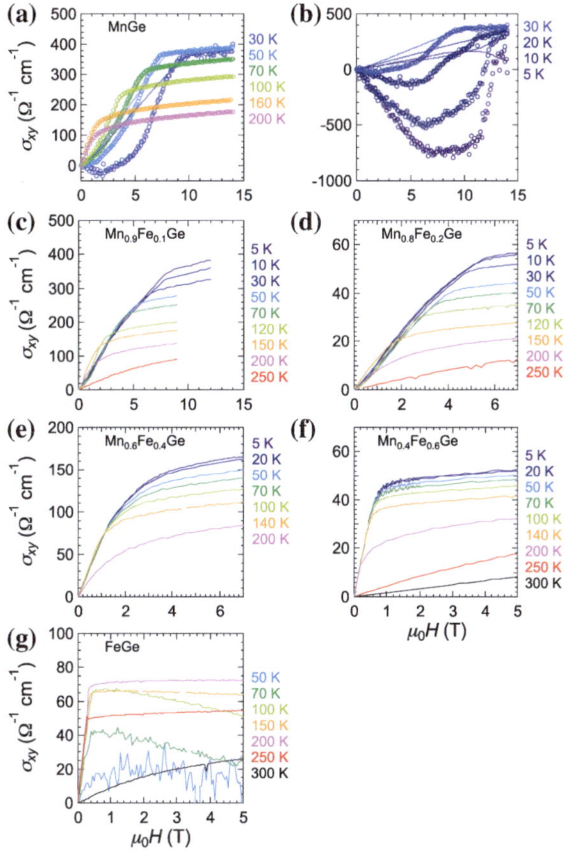

Fig. 3.13 Magnetic field dependences of Hall conductivities at various temperatures in $Mn_{1-x}Fe_xGe$. Except for MnGe and FeGe, the profiles resemble those of magnetizations

does not originate from being short-period helical structure. We attribute it to the skyrmion formation and discuss in detail in Chap. 4.

Figure 3.13 shows magnetic-field dependences of Hall conductivities $\sigma_{xy} = \frac{\rho_{yx}}{\rho_{xx}^2+\rho_{yx}^2}$ in $Mn_{1-x}Fe_xGe$. The fitting curves that $\sigma_{xy} = R_0\rho_{xx}^{-2}H + S_H M$ are only shown for the Hall conductivities in MnGe (Fig. 3.13a, b). In particular, the Hall conductivities of MnGe and $Mn_{0.9}Fe_{0.1}Ge$ exhibit large values up to $400\,\Omega^{-1}\,cm^{-1}$, which are much larger than those of MnSi ($\approx 200\,\Omega^{-1}\,cm^{-1}$) and $Fe_{1-x}Co_xSi$ (10–$40\,\Omega^{-1}\,cm^{-1}$).

Except for MnGe and FeGe, The temperature variations of σ_{xy}-H curves resemble those of M-H curves. We plot in Fig. 3.14 σ_{xy} and M at high fields as a function of T. As expected, σ_{xy} varies according to the variation of M in $Mn_{1-x}Fe_xGe$ ($0.1 \leq x \leq 0.6$). In contrast, σ_{xy} in MnGe and FeGe show considerable reductions in the low temperature regions, which may be due to skew-scattering effect appearing along with the enhancement of the mean free path [17].

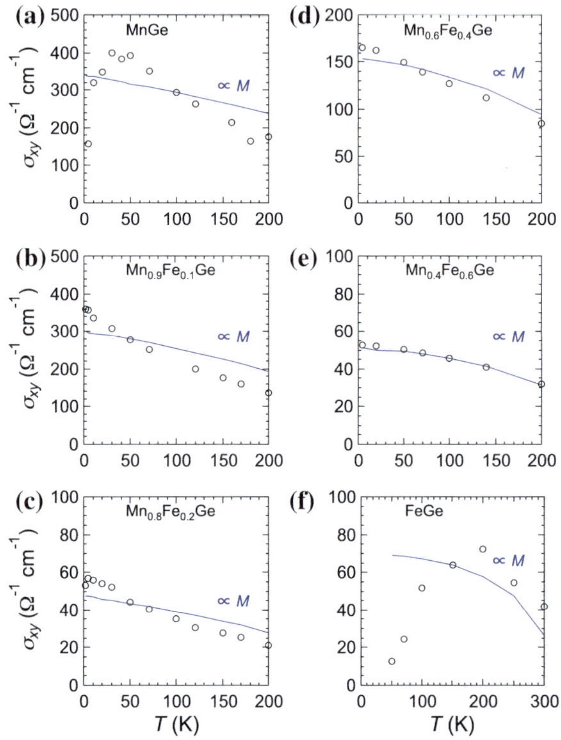

Fig. 3.14 Temperature dependences of Hall conductivities in $Mn_{1-x}Fe_xGe$, compared with their magnetizations

Temperature dependence of the anomalous Hall coefficient S_H is shown in Fig. 3.15. The values of S_H are almost equal to σ_{xy}/M, which indicates that the anomalous Hall effect is dominant in the Hall conductivity at the ferromagnetic phase. The temperature independent profile of S_H is again highlighted in $Mn_{1-x}Fe_xGe$ ($0.1 \leq x \leq 0.6$), while the constancy of S_H is true for high temperature ranges in MnGe and FeGe.

We plot the composition x dependence of the anomalous Hall coefficient S_H in Fig. 3.16. Here we take the values at 50 K, above which the Berry-phase scenario is also reasonable for MnGe and FeGe. The variation of S_H is reminiscent of that of the helical period observed in LTEM study [9]; we compare S_H and the strength of D [$\propto T_N/(\lambda M_s^2)$]. As the saturated magnetization M_s show a twofold change along the electron filling x, we make a small modification to estimate exchange interaction J: $T_N \propto JM^2$. A good agreement between S_H and D demonstrate the proportionality between the anomalous Hall effect and the spin-orbit interaction. In some compounds such as $SrRuO_3$ [18], the spin-orbit interaction cannot be perturbatively treated; the monopole structure at the band-crossing point in the momentum space induces a nontrivial temperature dependence of Hall signals. In contrast, in $Mn_{1-x}Fe_xGe$, the effective spin-orbit interaction to the AHE would be written by a perturbative treatment. $Mn_{1-x}Fe_xGe$ system is a good example demonstrating the spin-orbit interaction effect in the AHE.

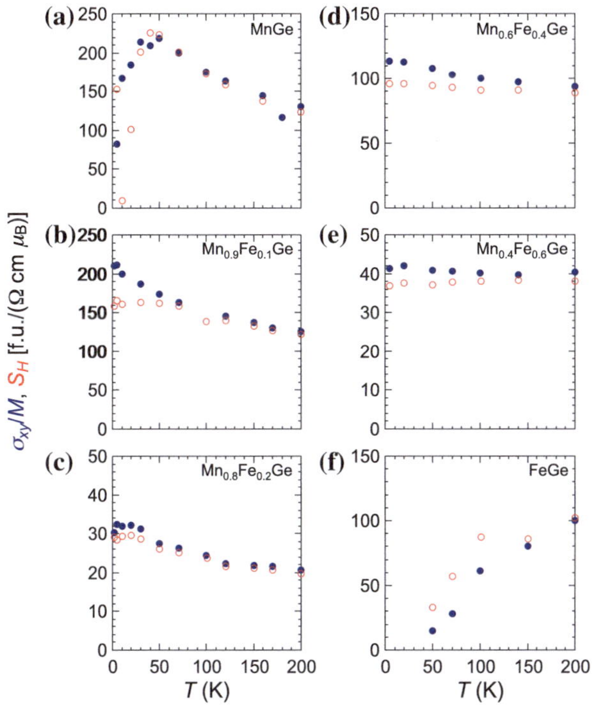

Fig. 3.15 Temperature dependences of σ_{xy}/M and $S_{\rm H}$ in $Mn_{1-x}Fe_xGe$, both of which represent anomalous Hall coefficients

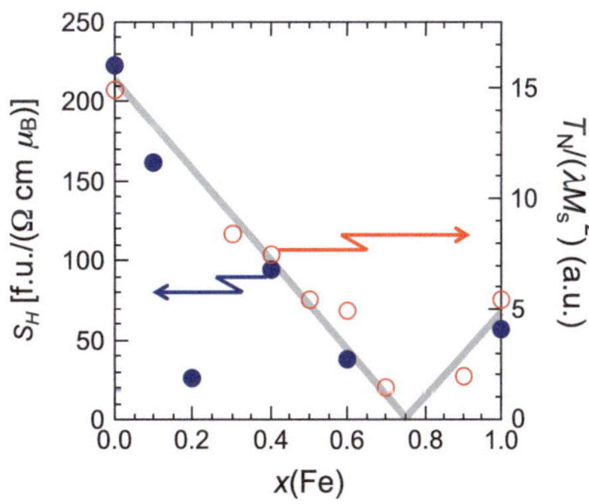

Fig. 3.16 Comparison between the anomalous Hall coefficient $S_{\rm H}$ and the strength of the DM interaction. Both the two composition-dependences fall into one curve indicated by a *thick gray line*

3.4 Summary

In this chapter, we have systematically investigated magnetization and Hall-effect profiles in a series of B20-type compounds $Mn_{1-x}Fe_xGe$ and $Fe_{1-y}Co_yGe$. Those compounds are categorized into the three magnetic phases: short-period helical magnet with strong spin-orbit interaction [$Mn_{1-x}Fe_xGe$ ($0 \leq x \leq 0.3$)], typical helical magnet similar to B20-type silicide magnets [$Mn_{1-x}Fe_xGe$ ($0.4 \leq x \leq 1$), $Fe_{1-y}Co_yGe$ ($0 \leq x \leq 0.84$)], and paramagnet (CoGe). In $Mn_{1-x}Fe_xGe$ ($0 \leq x \leq 0.3$), effects of strong spin-orbit coupling appear in spin-dependent transport properties, i.e., large magnetoresistance effect and large anomalous Hall current. Furthermore, a clear signal other than normal and anomalous Hall effects is observed in MnGe, which we attribute to formation of a non-coplanar magnetic structure below H_c. Anomalous Hall coefficient scales the effective spin-orbit interaction in all the composition range of $Mn_{1-x}Fe_xGe$, which demonstrates that $Mn_{1-x}Fe_xGe$ system is an excellent laboratory for the AHE due to Berry phase in momentum space.

References

1. N. Nagaosa, Y. Tokura, Phys. Scr. **T146**, 014020 (2012)
2. E.H. Hall, Philos. Mag. **10**, 301 (1880)
3. N. Nagaosa, J. Sinova, S. Onoda, A.H. MacDonald, N.P. Ong, Rev. Mod. Phys. **82**, 1539 (2010)
4. D. Xiao, M.C. Chang, Q. Niu, Rev. Mod. Phys. **82**, 1959 (2010)
5. M.V. Berry, Proc. R. Soc. Lond. A Math. Phys. Sci. **392**, 45 (1984)
6. Y. Onose, N. Takeshita, C. Terakura, H. Takagi, Y. Tokura, Phys. Rev. B **72**, 224431 (2005)
7. M. Lee, Y. Onose, Y. Tokura, N.P. Ong, Phys. Rev. B **75**, 172403 (2007)
8. M. Kataoka, O. Nakanishi, J. Phys. Soc. Jpn. **50**, 3888 (1981)
9. K. Shibata, X.Z. Yu, T. Hara, D. Morikawa, N. Kanazawa, K. Kimoto, S. Ishiwata, Y. Matsui, Y. Tokura, Nat. Nanotechnol. **8**, 723 (2013)
10. A. Bauer, A. Neubauer, C. Franz, W. Münzer, M. Garst, C. Pfleiderer, Phys. Rev. B **82**, 064404 (2010)
11. M.G. Chaudron, C.R. Acad, Sci. Paris Ser. C **281**, 527 (1975)
12. N. Kanazawa, Y. Onose, T. Arima, D. Okuyama, K. Ohoyama, S. Wakimoto, K. Kakurai, S. Ishiwata, Y. Tokura, Phys. Rev. Lett. **106**, 156603 (2011)
13. N. Kanazawa, J.-H. Kim, D.S. Inosov, J.S. White, N. Egetenmeyer, J.L. Gavilano, S. Ishiwata, Y. Onose, T. Arima, B. Keimer, Y. Tokura, Phys. Rev. B **86**, 134425 (2012)
14. P. Bak, M.H. Jensen, J. Phys. C **13**, L881 (1980)
15. O. Nakanishi, A. Yanase, A. Hasegawa, M. Kataoka, Solid State Commun. **35**, 995 (1980)
16. M. Kataoka, J. Phys. Soc. Jpn. **56**, 3635 (1986)
17. S. Onoda, N. Sugimoto, N. Nagaosa, Phys. Rev. B **77**, 165103 (2008)
18. Z. Fang, N. Nagaosa, K.S. Takahashi, A. Asamitsu, R. Mathieu, T. Ogasawara, H. Yamada, M. Kawasaki, Y. Tokura, K. Terakura, Science **302**, 92 (2003)

Chapter 4
3D Skyrmion-Lattice and Topological Hall Effect in MnGe

4.1 Introduction

4.1.1 Topological Hall Effect

Emergent electromagnetic fields originating from Berry phase [1, 2] can induce unique Hall responses as detailed in Chap. 3. In the case of anomalous Hall effect (AHE) in magnets, Berry phase in the momentum space plays the role causing scattering-free Hall current. A finite Berry phase can exist in any parameter space, and what we focus on in this Chapter is Berry phase defined in the real-space, resulting in so-called "topological Hall effect" (THE). The definition of THE seems under re-evaluation because a unified description of Hall effect in terms of Berry phase is now building consensus. Here we take a narrow definition that the topological Hall effect is induced by an emergent magnetic field originating from a "non-coplanar" magnetic texture. Conduction electrons in a non-coplanar spin structure can acquire the Berry phase Φ related to the spacial variation of the spin direction [3, 4]. In the strong coupling limit, i.e., the spin of the conduction electron is fully aligned along the underlying magnetic structure, the transfer integral in hopping from site i to j is calculated as

$$|\chi_i\rangle = \begin{pmatrix} \cos(\theta_i/2) \\ e^{i\phi_i}\sin(\theta_i/2) \end{pmatrix}, \tag{4.1}$$

$$t_{ij} = t\langle\chi_j|\chi_i\rangle = t\left[\cos\frac{\theta_i}{2}\cos\frac{\theta_j}{2} + \sin\frac{\theta_i}{2}\sin\frac{\theta_j}{2}e^{i(\phi_j-\phi_i)}\right], \tag{4.2}$$

where $|\chi_i\rangle$ is the two component spinor spin wave function, θ_i and ϕ_i are polar and azimuthal angles of a conduction spin, respectively. The phase factor is multiplied according to the tilting angles of spins. Then a Berry phase defined in a close loop consisting of three spins is calculated as $\mathbf{n}_i \cdot (\mathbf{n}_j \times \mathbf{n}_k) \propto \Omega/2$, where $\mathbf{n}_i$ is the unit vector of the spin of the conduction electron and Ω is the solid angle subtended by

N. Kanazawa, *Charge and Heat Transport Phenomena in Electronic and Spin Structures in B20-type Compounds*, Springer Theses,
DOI 10.1007/978-4-431-55660-2_4

the three spins. This product of three spins is referred to as "scalar spin chirality", which becomes a finite value if an arrangement of spins is a non-coplanar structure.

The effect of Berry phase due to spin chirality is, however, frequently canceled out by the sum over the whole lattice sites in real materials. Onoda et al. [5] pointed out that there are two mechanisms to avoid such cancellation and to induce THE. One is caused by the presence of inequivalent multiple loops in a unit cell. In this case, even if the total spin chirality is zero, the electrons acquire the Berry phase in the course of momentum change. The topological Hall effect induced by this mechanism has been observed for pyrochlores $Nd_2Mo_2O_7$ [6] and $Pr_2Ir_2O_7$ [7], which certainly have two inequivalent loops (hexagons and triangles). The other mechanism is due to the spin texture hosting the spin chirality, whose size is much larger than the lattice constant. One such example is a topological spin texture called skyrmion, in which the varying direction of the spins wraps a sphere. In the continuum limit, the total spin chirality is proportional to the number of the skyrmion. The electron moving over the skyrmion spin texture acquires the Berry phase proportional to the skyrmion density. The momentum-space emergent magnetic field $\mathbf{b_k}$ in the former case gives rise to the so-called anomalous velocity $\mathbf{v} = \frac{e}{\hbar}\mathbf{E} \times \mathbf{b_k}$. In this case, Hall conductivity is expressed as $\sigma_{xy} = \frac{e^2}{\hbar}\Sigma f(\varepsilon) b_{kz}$ [$f(\varepsilon)$ is Fermi distribution function], being independent of the scattering rate $1/\tau$ in a similar way to AHE proportional to magnetization: In ferromagnets, the spin-orbit interaction also gives rise to $\mathbf{b_k}$ and induces the anomalous Hall effect (AHE) proportional to the magnetization as introduced in Sect. 3.1. By contrast, the real-space emergent magnetic field $\mathbf{b_r}$ in the latter case is expected to give rise to the fictitious Lorentz force $F = e\mathbf{v} \times \mathbf{b_r}$ and Hall resistivity $\rho_{yx} = R_0 b_{rz}$, where the normal Hall coefficient R_0 is independent of $1/\tau$. Topological Hall effect due to the real-space emergent field has a chance to appear in a topological spin texture.

In a skyrmion magnetic structure, the Berry phase collected by the conduction electrons is calculated as

$$\Phi = 2\pi \int d\mathbf{r}\boldsymbol{\Phi}^z = \frac{1}{2}\int d\mathbf{r}\, \mathbf{n} \cdot \left(\frac{\partial \mathbf{n}}{\partial x} \times \frac{\partial \mathbf{n}}{\partial y}\right) = \frac{\Omega}{2}, \tag{4.3}$$

which can result in the topological Hall effect; the topological Hall resistivity ρ_{yx}^{T} is in proportion to Φ. Here $\boldsymbol{\Phi}^z$ is the skyrmion density, and the integration extends over one magnetic unit cell. The calculation of the Berry phase in the skyrmion system is essentially equal to counting the number of skyrmions because one skyrmion subtends a solid angle $\Omega = 4\pi$ (a sphere).

In analogy with Aharonov-Bohm effect, the Berry phase is treated as the gauge flux $\frac{\phi_0}{2\pi}\Phi$, where ϕ_0 is the magnetic flux quantum h/e. The Aharonov-Bohm effect describes that an electron moving along a close loop, in which a magnetic flux $\phi = BA$ pierces, collects a phase $2\pi\frac{\phi}{\phi_0}$. Here B is a magnetic field and A is an area of the closed loop. And thus, the emergent magnetic field b_{rz} due to the Berry phase Φ can be calculated to be $-\frac{\phi_0}{2\pi}\frac{\Phi}{A}$. In the case of the triangular lattice of skyrmions, the nearly uniform emergent magnetic field is

$$b_{rz} = -\frac{\sqrt{3}}{2\lambda^2}\phi_0. \tag{4.4}$$

The minus sign of this relation originates from the fact that the core magnetization of the magnetic skyrmion in B20-type compounds is always anti-parallel to the magnetic field (z-axis).

A topological Hall effect, assigned to the outcome of SkX, has already been observed in MnSi at ambient pressure [8] and high pressure [9]. However, the Hall anomaly remains very small ($|\rho_{yx}^{\mathrm{T}}| \approx 4$ nΩ cm) and appears in a very narrow temperature region ($T = 28 - 29$ K) at ambient pressure (Fig. 4.1); the nature of the THE, such as the $1/\tau$ dependence, remains veiled.

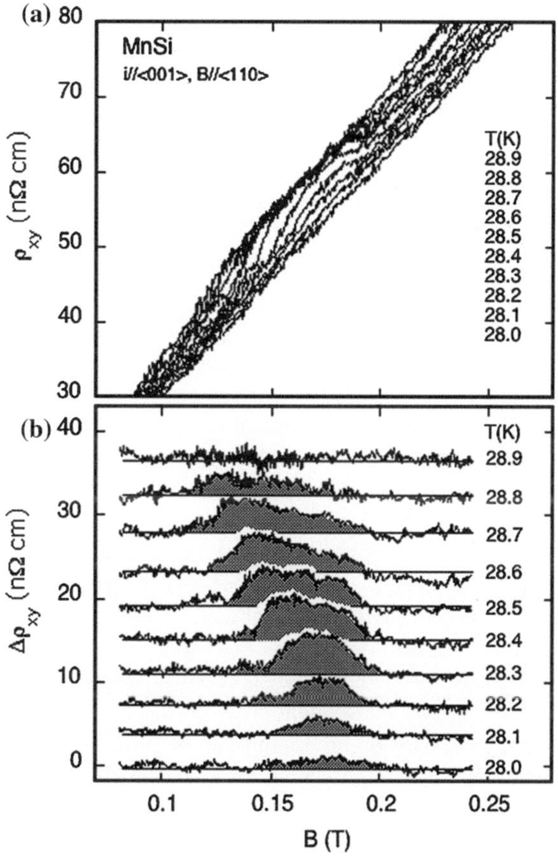

Fig. 4.1 Topological Hall effect in bulk MnSi at ambient pressure [8]. Topological Hall anomaly is observed at skyrmion phase just below T_N. Reprinted with permission from Ref. [8]. Copyright 2009 by American Physical Society

4.2 Topological Hall Effect in MnGe

We have studied another B20-type material MnGe, where a short period helical spin structure is expected from the magnetization measurements shown in Chap. 3. A discernible Hall response caused by the non-coplanar spin background is observed in a much wider temperature-field region than MnSi. Importantly, we have found that the topological Hall resistivity is independent of the scattering rate. This indicates the topological Hall effect is caused by the Berry phase of electrons acquired in the course of the real-space motion in contrastive with the momentum-space Berry phase for anomalous Hall effect in ferromagnets.

First we overview the basic properties of MnGe. Figure 4.2a shows the T-dependence of resistivity and susceptibility of MnGe. The resistivity exhibits a metallic behavior with the residual resistance ratio $\rho(300\text{ K})/\rho(2\text{ K}) \approx 19$, while the susceptibility shows a broad antiferromagnetic peak around $T_{\rm N} \approx 170$ K. Shown in the inset of Fig. 4.2b are high-angle neutron diffraction patterns around (1 1 0) reflection at various temperatures as subtracted by the profile at 270 K to highlight magnetic reflections. Two peaks of magnetic origin were discerned at each temperature below $T_{\rm N}$. The integrated intensity of these reflections decreases with increasing temperature and finally disappears above $T_{\rm N}$. Two and four magnetic peaks also appear around (1 1 1) and (2 1 0) reflections, respectively. Because a magnetic reflection, in general, appears at $|\mathbf{q}| = |\mathbf{Q}_l \pm \mathbf{Q}_m|$ in a powder neutron diffraction pattern, these magnetic peak positions can be accounted for with an incommensurate helical magnetic structure with $\mathbf{Q}_m = \{\delta\ 0\ 0\}$; the magnetic peaks around (1 1 0), (1 1 1), and (2 1 0) reflections are indexed as $q = \{1 \pm \delta\ 1\ 0\}$, $q = \{1 \pm \delta\ 1\ 1\}$, and $q = \{2 \pm \delta\ 1\ 0\}$, $\{1 \pm \delta\ 2\ 0\}$, respectively. Here we define $\mathbf{q}$, $\mathbf{Q}_l$, and $\mathbf{Q}_m$ as the scattering vector, the reciprocal lattice vector, and the magnetic modulation vector, respectively. The helical period varies from 6 to 3 nm with decreasing temperature, which is the shortest among the hitherto known B20-type helimagnets (Fig. 4.2c), promising the highest-density skyrmions.

Figure 4.3 presents magnetic-field dependences of magnetization M, magnetoresistivity $\rho_{xx}(H)/\rho_{xx}(0)$, and Hall resistivity ρ_{yx}. The magnetization curve shows a kink representing the full alignment of the spins parallel to the applied field, which we define as the critical field $H_{\rm c}$. Much larger $H_{\rm c}$ in MnGe (≈12 T at 5 K) than that in other B20-type helimagnets (below 1 T) indicates strong Dzyaloshinsky-Moriya interaction and consequent short-period magnetic structure as detailed in Chap. 3. Now, let's focus on the main issue of Hall resistivity data (Fig. 4.3c, d). Especially below 30 K, the sign change of Hall resistivity occurs, which clearly demonstrate an additional component other than anomalous Hall effect because the magnetization increases linearly proportional to the magnetic field below $H_{\rm c}$ (Fig. 4.3a). The Hall anomaly, which deviates from the shape of the magnetization curve, becomes increasingly conspicuous with decreasing temperature. The additional component of the Hall effect produces a negative value of ρ_{yx} besides AHE and H-linear normal Hall effect (NHE), which we attribute to THE associated with the skyrmion formation. To estimate this additional part $\rho_{yx}^{\rm T}$, we assume the relation,

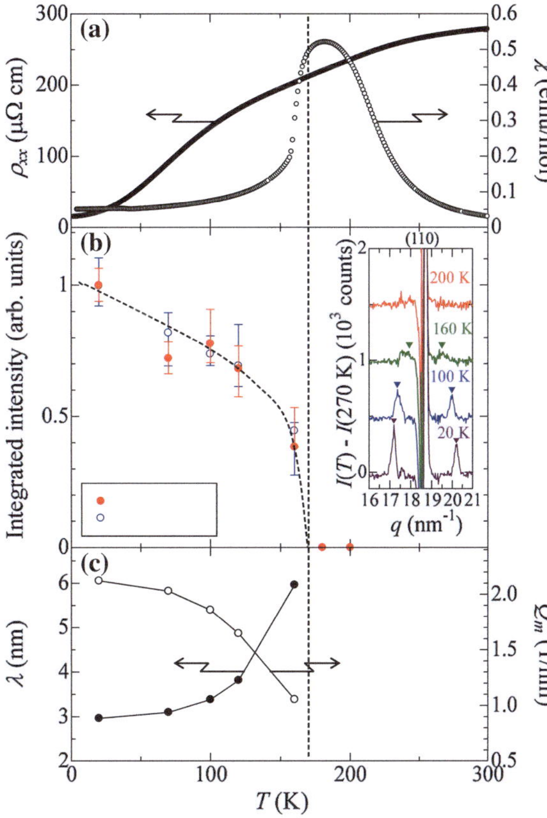

Fig. 4.2 **a** Temperature dependence of resistivity ρ_{xx} and magnetic susceptibility ($\chi = M/H$ at $H = 100$ Oe) in MnGe. **b** Temperature dependence of the integrated intensity of magnetic Bragg peaks indexed with the propagation vector $\mathbf{Q}_m = \{\delta\ 0\ 0\}$. The integrated intensity is normalized by that at 20 K. Inset shows neutron powder diffraction patterns of 20, 100, 160, and 200 K as subtracted by that at 270 K > $T_{\rm N}$. **c** Temperature dependence of the helical period λ and the magnitude of $\mathbf{Q}_m$. The *dashed* and *solid lines* in (**b**) and (**c**) are the guide to the eyes. Reprinted with permission from Ref. [10]. Copyright 2011 by American Physical Society

$$\begin{aligned}\rho_{yx} &= \rho_{yx}^{\rm N} + \rho_{yx}^{\rm A} + \rho_{yx}^{\rm T} \\ &= R_0 H + S_{\rm H}\rho_{xx}^2 M + \rho_{yx}^{\rm T}. \end{aligned} \tag{4.5}$$

Here, $S_{\rm H}$ is an H-independent parameter of AHE [11]. This relation is also based on the additivity of resistivity, which is satisfied when $\rho_{yx} \ll \rho_{xx}$, because the relation originally comes from the additivity of Hall currents. As the topological Hall resistivity should vanish when spins become fully polarized at $H > H_{\rm c}$, we can determine R_0 and $S_{\rm H}$ as the slope and the intercept of the curve ρ_{yx}/H versus $\rho_{xx}^2 M/H$ above the critical field $H_{\rm c}$. As shown in Fig. 4.3c, d, we obtained a good fit above $H_{\rm c}$ with the fitting parameters R_0 and $S_{\rm H}$; above $H_{\rm c}$, the magnetoresistance (a decrease of ρ_{xx}, Fig. 4.3b) is important to explain a decrease of Hall resistivity through the anomalous Hall term $\propto S_{\rm H}\rho_{xx}^2$. As a result of the fitting, the additional contribution $\rho_{yx}^{\rm T}$ is estimated as the subtraction of $\rho_{yx}^{\rm N} + \rho_{yx}^{\rm A}$ (solid lines in Fig. 4.3c, d) from measured ρ_{yx} (open circles in Fig. 4.3c, d). Note that a peak shape of $\rho_{yx}^{\rm T}$ around $H_{\rm C}$ above 50 K is perhaps not essential but merely a byproduct; this is due to

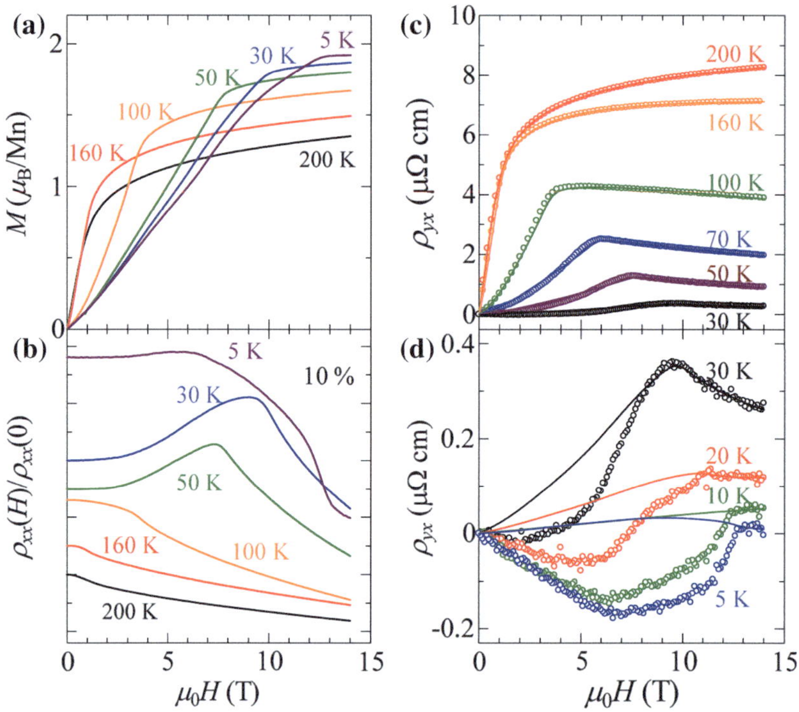

Fig. 4.3 Magnetic-field dependence of **a** magnetization M, **b** resistivity ρ_{xx} as normalized by the zero-field value $\rho_{xx}(0)$, and **c**, **d** Hall resistivity ρ_{yx} at various temperatures. *Solid lines* in panels (**c**) and (**d**) are the fitted curves of ρ_{yx} using the relation that $\rho_{yx} = R_0 B + S_H \rho_{xx}^2 M$ with the fitting parameters R_0 and S_H, to represent the conventional normal Hall and anomalous Hall components, respectively (see text). Reprinted with permission from Ref. [10]. Copyright 2011 by American Physical Society

difficulty in estimate of the small ρ_{yx}^T buried in the much larger ρ_{yx}^A, which increases in proportion to ρ_{xx}^2 with increasing temperature. Nevertheless, we can obtain valid estimation below 70 K as is evident also from the raw data shown in Fig. 4.3d. We find almost constant quantity, $\rho_{yx}^T \approx -0.16\,\mu\Omega$ cm, against temperature variation below 70 K.[1] This temperature independence is the key for the topological feature of ρ_{yx}^T. We show the T-dependence of ρ_{yx}^A (σ_{xy}^A) and ρ_{yx}^T (σ_{xy}^T) in Fig. 4.4a, b, where we plot ρ_{yx}(14 T) [σ_{xy}(14 T)] as ρ_{yx}^A (σ_{xy}^A) and the negative peak value of ρ_{yx}^T (σ_{xy}^T), respectively. Both ρ_{yx}^A and ρ_{yx}^T are in good agreement with the anticipated relations, $\rho_{yx}^A = S_H \rho_{xx}^2 M \propto M/\tau^2$ and $\rho_{yx}^T = R_0 b_{rz} \propto \tau^0$, respectively. This again ensures that ρ_{yx}^T is induced by totally different mechanism from that of the M-linear AHE,

[1]The magnetic modulation period is almost constant below 70 K; the emergent field also remains constant in the temperature range of the present focus.

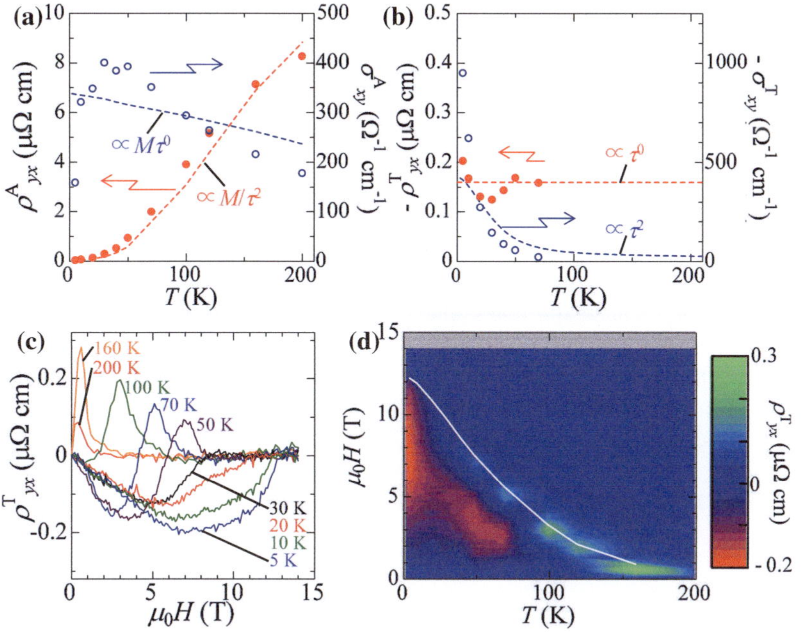

Fig. 4.4 **a**, **b** Comparison between T-dependence of **a** anomalous and **b** topological components of Hall resistivity (*solid circles*) and conductivity (*open circles*). ρ_{yx}^{A} is here defined by the value of $\rho_{yx}(14\ \mathrm{T})$, while ρ_{yx}^{T} here by the negative maximal values of ρ_{yx}^{T}-H curves in panel (**c**). **c** Magnetic-field dependence of topological Hall resistivity ρ_{yx}^{T}, which is deduced by the subtraction of the conventional anomalous plus normal Hall resistivities $\rho_{yx} = R_0 H + S_{\mathrm{H}} \rho_{xx}^2 M$ from the measured ρ_{yx}. **d** A contour map of ρ_{yx}^{T} in the plane of temperature and magnetic field. The *white curve* represents the temperature variation of the critical field H_{c}, at which the spins are fully aligned. Reprinted with permission from Ref. [10]. Copyright 2011 by American Physical Society

supporting our conclusion that ρ_{yx}^{T} originates in the effective field action due to the scalar spin chirality. The nearly T-independent behavior of ρ_{yx}^{T} can also be clearly discerned in the contour mapping (Fig. 4.4d). The absolute magnitude of the observed ρ_{yx}^{T} ($\approx 0.16\,\mu\Omega$ cm) for MnGe is 40 times as large as that for MnSi ($\approx 0.004\,\mu\Omega$ cm [8]). This ratio appears quite reasonable; the magnetic modulation period differs in the two materials, $\lambda = 3$ nm for MnGe and $\lambda = 18$ nm for MnSi, leading the difference in the possible total spin-chirality (e.g. skyrmion density) as much as $(18/3)^2 = 36$ times.

There is, however, a distinct field dependence of ρ_{yx}^{T} in MnGe, which is not consistent with those of the hitherto known skyrmionic B20 materials (see also Figs. 4.1 and 4.4c): ρ_{yx}^{T} begins to appear immediately upon the application of the magnetic field and persists up to the critical field H_{c} above which spins are fully aligned. Furthermore, the temperature region with discernible ρ_{yx}^{T} is very wide from

2 to 70 K. From small-angle neutron scattering experiments and calculations of Berry phase, we can provide a description of the peculiar topological Hall effect, i.e., predicting a model of formation of cubic lattice of skyrmions.[2]

4.3 Small-Angle Neutron Scattering on MnGe—Possible 3D Skyrmion Lattice

We discuss the form of the non-coplanar magnetic texture producing the distinct topological Hall effect in MnGe based on results of small-angle neutron scattering (SANS) experiments on MnGe powder sample as well as a theoretical analysis of ρ_{yx}^{T} [16]. Emergence of magnetic modulations perpendicular to an applied magnetic field indicates a formation of a multiple-**Q** magnetic structure, or a skyrmion crystal state. Either a square or a simple-cubic skyrmion lattice shows good agreements with SANS intensity patterns and topological Hall signals.

4.3.1 Small-Angle Neutron Scattering on MnGe

The SANS patterns during zero field cooling (ZFC) from room temperature are shown in Fig. 4.5a–f. A diffusive scattering intensity around $q = 0$ gradually grows with decreasing temperature above $T_{\mathrm{N}} \approx 170$ K, and Debye-ring like patterns are clearly observed below T_{N} as evidenced by the angle-averaged diffracted intensity as a function of wavevector in Fig. 4.5g. The enhancement of the intensity around $q = 0$ indicates that the magnetic correlation is developing well above T_{N}. On the other hand, the ring patterns below T_{N} are associated with the formation of a periodically modulated magnetic structure and the radius of the diffracted neutron pattern, i.e., the magnitude of the magnetic modulation vector, increases with decreasing temperature from 1.15 nm^{-1} (151 K) to 2.23 nm^{-1} (30 K), which is in accord with previous reports (Fig. 4.5h) [10, 17].

Figure 4.6 shows the development of SANS patterns under different magnetic fields at 30 K after ZFC. We could not access the grey shaded region in Fig. 4.6 due to the angular limitation of the cryomagnet window. When applying a field of 2 T, three characteristic diffraction patterns appear: crescent-shaped intensity patterns parallel to the field $I(\mathbf{q} \parallel \mathbf{H})$ (patterns 1 and 2 in Fig. 4.6b), a local maximal intensity pattern perpendicular to the field $I(\mathbf{q} \perp \mathbf{H})$ (intensity 3 in Fig. 4.6b), and a diffusive pattern around the center I_{diff} (intensity 4 in Fig. 4.6b). At higher magnetic fields of 4 and 6 T (Fig. 4.6c, d), $I(\mathbf{q} \perp \mathbf{H})$ almost disappears and the crescent shape of $I(\mathbf{q} \parallel \mathbf{H})$ shrinks. Around the critical field H_{c} ($\approx$9.8 T at 30 K), all the intensities sharply drop (Fig. 4.6e). When ramping the field down back to zero, the various

[2]Various skyrmion ground states other than the triangular lattice form have been theoretically proposed [12–15].

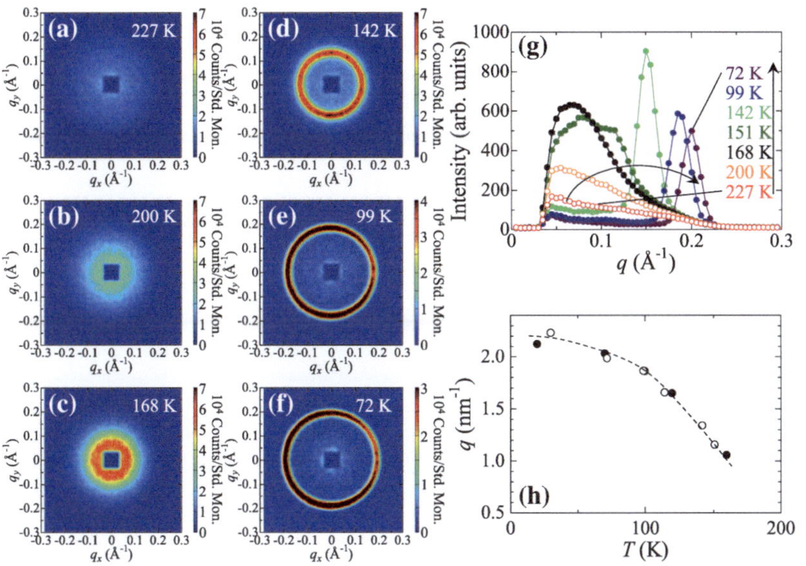

Fig. 4.5 **a–f** SANS patterns at zero magnetic field during zero-magnetic-field cooling (ZFC). **g** The angle-averaged SANS intensity during ZFC as a function of wavevector q. **h** Temperature dependence of the magnetic modulation vector (*open circles*) as measured in the present SANS study, compared with the data by the high-angle neutron diffraction study reproduced from Ref. [10] (*solid circles*). Reprinted with permission from Ref. [16]. Copyright 2012 by American Physical Society

overall SANS profiles at each magnetic field (Fig. 4.6f, g) are similar to those which were measured on increasing the field (Fig. 4.6b, c), although the ring pattern is not restored and all the characteristic intensities remain at zero magnetic field (Fig. 4.6h). Rough alignment of the modulation directions parallel to the field is reflected to the crescent shape of $I(\mathbf{q} \parallel \mathbf{H})$, which is also detected in a single crystalline B20-type $Fe_{0.7}Co_{0.3}Si$ (Ref. [18]). The shrinkage of the crescent shape indicates the gradual alignment of modulation vectors toward $\mathbf{q} \parallel \mathbf{H}$. I_{diff} (intensity 4) is assigned to the double scattering of the helical magnetic structure, representing the superposition of two crescent shapes of intensities 1 and 2. The emergence of intensity perpendicular to the applied field is a typical signature of the formation of a skyrmion lattice, which has been confirmed for the skyrmion lattice state in other B20-type compounds [18, 19].

Here we should note that the magnetic field causes partial alignment of the MnGe powder grains as well. Once a magnetic field is applied to the sample, the ring-shaped SANS pattern below T_N gives way to an anisotropic crescent-shaped one as shown in Fig. 4.6h. The identical change happens even after initializing (randomizing) the magnetic state by increasing the sample temperature well above T_N (e.g. 300 K, not shown). This demonstrates that the easy axes of magnetization, which are $\langle 100 \rangle$ in MnGe [10], are flipped along the field direction. Therefore, the peak intensities

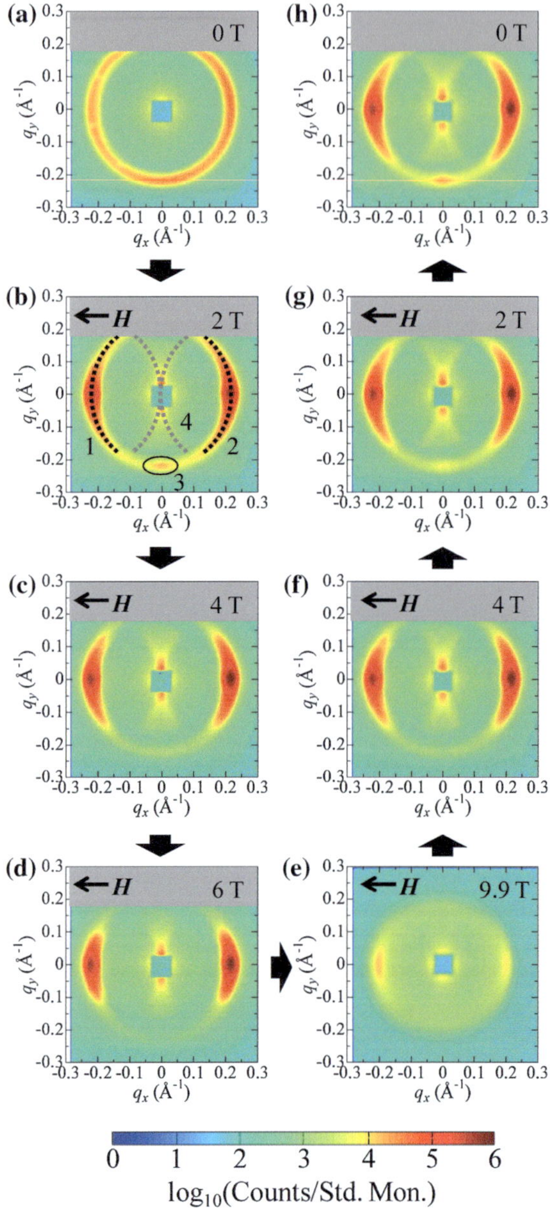

Fig. 4.6 Variation of SANS patterns with magnetic field (H) at 30 K. The magnitude of the applied magnetic field changes from (**a**) 0 T to (**e**) 9.9 T and back to (**h**) 0 T in the sequential order indicated by *thick arrows*. Three typical intensities are clearly observed in panel (**b**): (i) crescent-shaped diffractions parallel to the field (diffractions 1 and 2), (ii) a local maximal intensity observed perpendicular to the field (diffraction 3), which indicates the formation of a skyrmion lattice, (iii) a diffusive diffraction around the center (diffraction 4), which is assigned to the double scattering stemming from the crescent shape of diffractions 1 and 2. The *solid* and *dashed lines* in panel (**b**) are the guide to the eyes. Reprinted with permission from Ref. [16]. Copyright 2012 by American Physical Society

are observed again along the previous field direction, originating from the ⟨100⟩-modulated magnetic structure at zero field. Although the polycrystalline powder is subjected to the magnetic-field induced alignment, the assignment of the $I(\mathbf{q} \perp \mathbf{H})$ diffraction to the formation of a skyrmion lattice is supported by the following arguments: (i) If MnGe were in a single-q screw or conical state, the magnetic field would not be able to rotate the modulation directions perpendicular to its direction. (ii) Furthermore, although we cannot determine whether a single-q or the multiple-q state (skyrmion lattice) produce the intensity $I(\mathbf{q} \perp \mathbf{H})$ because of using the polycrystalline sample, the multiple-q state with the non-coplanar spin configuration is more plausible structure due to the existence of the topological Hall effect. (iii) The hysteresis behavior of the H-dependence of the magnetization in MnGe indicates that the modulation direction remains along the previously applied magnetic field once the field is increased above a critical value [10]. This feature is generically observed in other B20-type helimagnets, which is also confirmed by SANS experiments [18, 19]. Therefore, the reappearance of $I(\mathbf{q} \perp \mathbf{H})$ with decreasing field (Fig. 4.6g, h) could hardly be explained without assuming a formation of the skyrmion-lattice state, which has q vectors perpendicular to the magnetic field.

To summarize the above, a magnetic structure modulated in multiple directions, most probably a skyrmion lattice state, is stabilized as the ground state at zero magnetic field. Prior to the magnetic-field induced orientation of the polycrystalline powder, the randomly-oriented multiple-q structures produce the ring scattering pattern at zero magnetic field (Fig. 4.6a). The magnetic field (H) flips and aligns the modulation directions as well as the crystalline grains with the ⟨100⟩ axes parallel to H, giving rise to accumulated intensities parallel to the field. In the multiple-q state, the other modulation directions are perpendicular to the field, which can show up as $I(\mathbf{q} \perp \mathbf{H})$. The multiple q vectors aligned by the magnetic field persist and point along the same directions after removing the field, which results in a SANS pattern at zero field (Fig. 4.6h) similar to the one under a magnetic field (Fig. 4.6g). From the results of a high-angle neutron diffraction experiment [10], it is clear that all the multiple modulation directions should belong to the category ⟨100⟩, i.e. [100] and equivalent, directions. Therefore, the ground state will be the multiple-q state composed of independent two or three ⟨100⟩ directions, which we define as a square or a simple-cubic (SC) skyrmion lattice, respectively. If the square skyrmion lattice is the case, the observed much larger intensity $I(\mathbf{q} \parallel \mathbf{H})$ than $I(\mathbf{q} \perp \mathbf{H})$ may indicate that the skyrmion lattice coexists with the helical structure producing $I(\mathbf{q} \parallel \mathbf{H})$. On the other hand, in the case of the simple-cubic lattice, one modulation vector gets collected along the magnetic field and the set of the other two modulations can point to an arbitrary direction in the plane normal to the field, which may explain $I(\mathbf{q} \parallel \mathbf{H}) \gg I(\mathbf{q} \perp \mathbf{H})$.

Although some form of skyrmion lattice is likely to form in MnGe judging from the circumstantial evidences, i.e. the topological Hall effect and the appearance of $I(\mathbf{q} \perp \mathbf{H})$, there would be alternative scenarios to explain $I(\mathbf{q} \perp \mathbf{H})$ that involves only helices. For example, the grains may have cleavage surfaces of (100) and/or (110) planes [20], whose normal vectors are perpendicular to the ⟨100⟩ direction aligned along the field direction. If the Hamiltonian includes a term that align the helical plane with the surface of the powder grains, the additional spot $I(\mathbf{q} \perp \mathbf{H})$ in diffraction pattern may be explained. Nevertheless, those simple (single-q) helix structures can hardly be reconciled with the emergence of very large topological Hall effect as observed.

Hereafter, as the most plausible scenario, we adopt the standpoint that the skyrmion lattice is formed in MnGe. To identify the genuine magnetic contributions with $I(\mathbf{q} \parallel \mathbf{H})$ and $I(\mathbf{q} \perp \mathbf{H})$ or to eliminate the apparent variation of the scattering intensity by rotation of the grains, it is useful to discuss the H-dependence of the intensity in a decreasing field, since we can reasonably assume that the grains remain locked in their positions once they have been aligned by the field. In fact, two independent field-decreasing scans show little difference (open and closed data-points in Fig. 4.7a, b). The neutron scattering cross section is proportional to $|\hat{\mathbf{q}} \times [\mathbf{M}(\mathbf{q}) \times \hat{\mathbf{q}}]|^2$ [$\mathbf{M}(\mathbf{q})$ being the Fourier transform of the magnetization], and hence the scattering intensity coming from the conical magnetic structure is proportional to the squared magnetization component perpendicular to the scattering vector, $M_\perp^2 = M_s^2 - M^2$. M_s stands for the nearly-saturated magnetization e.g. at 14 T and M is the magnetization component parallel to the field. The H-dependence of $I(\mathbf{q} \parallel \mathbf{H})$ is proportional to $M_s^2 - M^2$. On the other hand, $I(\mathbf{q} \perp \mathbf{H})$ only rises below approximately 4 T, which can clearly be seen in the development of the peak intensity at $\theta = 0$ in the angular dependence of the scattering intensity (Fig. 4.7c). (Here, θ is defined as the angle of the q point in the q-circle measured from the bottom $q \perp H$ point.) Disarray of the

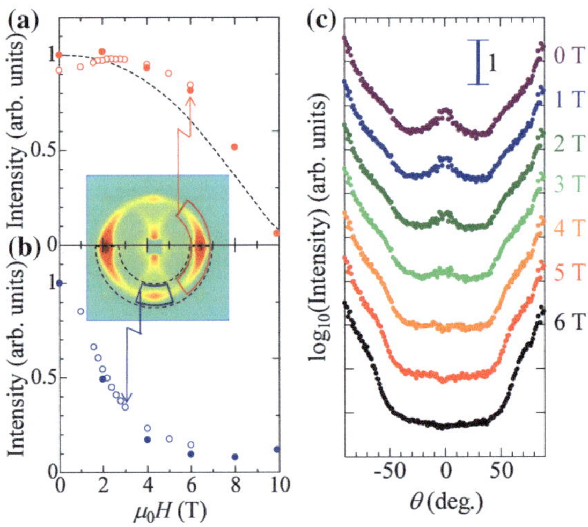

Fig. 4.7 Magnetic-field dependence of SANS intensities **a** parallel and **b** perpendicular to the field at 30 K. *Solid* and *open circles* represent different field-decreasing scans. The *dashed line* in panel (**a**) indicates the relation $M_s^2 - M^2$ versus H. **c** SANS intensity as a function of angle in the sector bounded by the *dashed line* in the inset of panels (**a**) and (**b**). The angle origin ($\theta = 0$) is defined at the $I(\mathbf{q} \perp \mathbf{H})$ position (the bottom of the *half circle*). Reprinted with permission from Ref. [16].

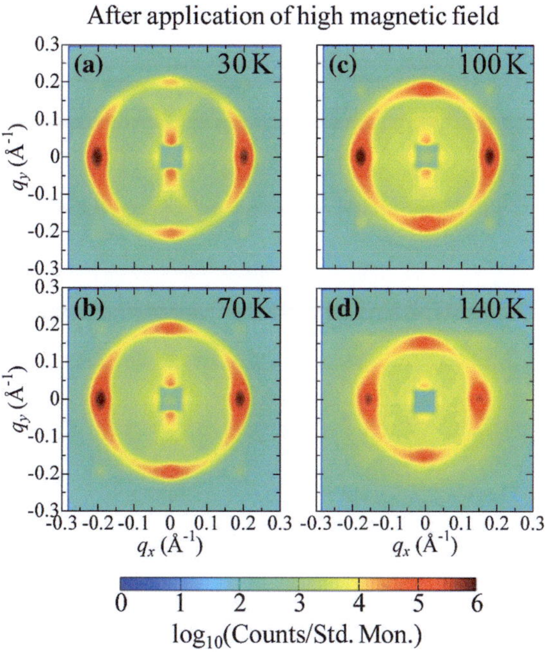

Fig. 4.8 SANS patterns under zero magnetic field after an application of high magnetic field exceeding critical magnetic field at each temperature; 9.9 T for 30, 70, 100 K and 5.0 T for 140 K. Reprinted with permission from Ref. [16]. Copyright 2012 by American Physical Society

skyrmion lattice appears to occur at high fields, which may explain the decrease of $I(\mathbf{q} \perp \mathbf{H})$ above 4 T; this is reminiscent of the transformation into a heavily distorted skyrmion lattice or sparsely populated skyrmions as observed in a $Fe_{0.5}Co_{0.5}Si$ thin plate [21] and has been explained theoretically [22].

We have also examined the temperature range where $I(\mathbf{q} \perp \mathbf{H})$ appears, i.e., the possible skyrmion lattice forms (Fig. 4.8). In order to exclude the possibility of a multi-domain state of single-q helical structure, SANS patterns were taken at zero magnetic field after aligning the modulation vector by applying a high magnetic field above H_c at each temperature ($\mu_0 H = 9.9$ T for 30, 70, 100 K and $\mu_0 H = 5.0$ T for 140 K). There still exists $I(\mathbf{q} \perp \mathbf{H})$ at all the measured temperatures. Thus, the skyrmion state is stable in a wide temperature range, perhaps over the whole temperature region below T_N (=170 K).

4.3.2 Fitting of Topological Hall Resistivity

The H-dependence of the topological Hall effect have been theoretically discussed for different forms of skyrmion lattices [14, 15, 23]. The profile is different from each other, being sensitive to the lattice forms. Here, to estimate ρ_{yx}^{T}, we follow the method by Binz and Vishwanath [23], and discuss which type of lattice, square or SC lattice, is more plausible for the case of MnGe. Assuming that the conduction electrons cannot couple to a part with smaller magnetization than some cutoff [23], the collected Berry phase is modified as $\Phi = 2\pi \int d\mathbf{r}\Theta(|\mathbf{M}(\mathbf{r})| - M_c)\Phi^z$, where Θ is the Heaviside step function and M_c is a cutoff magnetization.

The position dependence of the magnetization $\mathbf{M}(\mathbf{r})$ in the skyrmion lattice is described by a superposition of helices with identical handedness. The handedness is determined by the sign of the DM interaction. The square and SC skyrmion lattices are superpositions of two and three orthogonal helical structures, respectively;

$$\begin{aligned}\mathbf{M}^{\mathrm{sq.}}(\mathbf{r}) &= (0, \sin x, \cos x) + (-\sin y, 0, \cos y) + (0, 0, m) \\ &= (-\sin y, \sin x, \cos x + \cos y + m),\end{aligned}$$

and

$$\begin{aligned}\mathbf{M}^{\mathrm{SC}}(\mathbf{r}) &= (0, \cos x, -\sin x) + (-\sin y, 0, \cos y) + (\cos z, -\sin z, 0) + (0, 0, m) \\ &= (-\sin y + \cos z, -\sin z + \cos x, -\sin x + \cos y + m).\end{aligned}$$

Here, m is the induced magnetization which should be almost linearly proportional to the applied field as observed [10], The square and SC skyrmion lattices are illustrated in Fig. 4.9a, b, respectively, where m was set to $m = 0$. Arrays of vortices pointing up ($M_z > 0$; red) and down ($M_z < 0$; blue) are arranged in an alternate order for these skyrmion lattices. We can see nodes with zero M, in the vicinity of which we neglect the collection of the Berry phase. Note that the choice of chirality of skyrmion lattice, or equivalently the swirling direction in a skyrmion changes neither the value of the Berry phase nor the topological Hall effect; here we take the superpositions of left-handed spirals.

The calculated results of the Berry phase with different cutoffs are shown in Fig. 4.9d, e as a function of magnetic field in comparison with the observed topological Hall resistivity at 30 K in MnGe [10]. The negative maximal value of the Berry phase and the critical field in the calculations are fitted with the experimental negative peak value of ρ_{yx}^{T} and H_c, respectively. Both the results of the square and SC lattices explain crudely the experimentally observed profile of ρ_{yx}^{T}: (i) $\rho_{yx}^{\mathrm{T}} = 0$ at $H = 0$, (ii) a finite ρ_{yx}^{T} at $0 < H < H_c$, and (iii) $\rho_{yx}^{\mathrm{T}} = 0$ at $H > H_c$. These behaviors can be qualitatively explained as follows; (i) positive (skyrmion) and negative (anti-skyrmion) contributions cancel each other out at $H = 0$, (ii) finite m induced by a magnetic field breaks the balance of the contributions from skyrmions and anti-skyrmions, yielding nonzero net spin chirality, (iii) a fully spin polarized state gives no Berry phase factor. Among them, the striking agreement with the experimental data in the SC lattice model with the cutoff of $M_c = 1/6$ is noteworthy, while the square lattice model cannot give a good fit with any arbitrary cutoff value.

For comparison, we show in Fig. 4.9f the calculations of the Berry phase in the hitherto-known skyrmion lattice, i.e., the triangular lattice of anti-skyrmions illustrated in Fig. 4.9c. Because the triangular lattice is composed entirely of anti-skyrmions producing a negative fictitious field, a finite topological Hall effect would be observed even at zero magnetic field [15], which is discrepant with the observation.

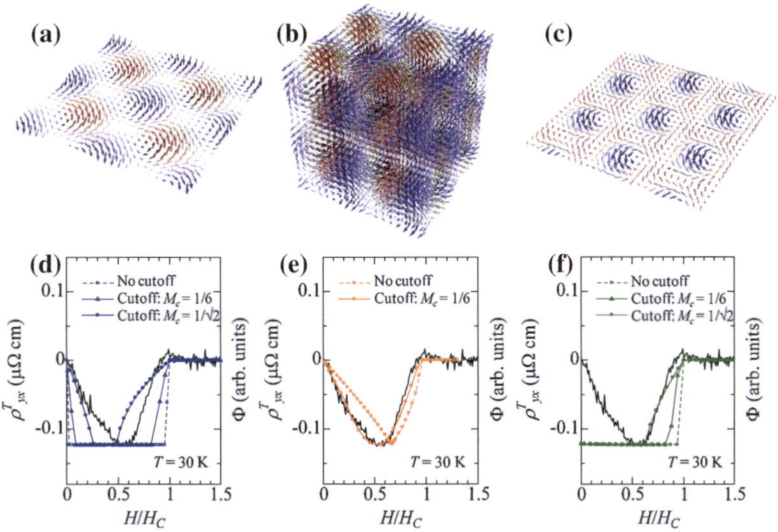

Fig. 4.9 Schematic illustrations of **a** a square, **b** a simple-cubic, and **c** a triangular skyrmion lattice. The *color* and the hue of the *arrows* represent the *z*-component of the magnetic moments; *red* (*blue*) *arrows* indicate up (down) magnetic moments. Comparison between the experimental results of topological Hall resistivity [10] (*black solid lines*) and the calculated ones with varying cutoff magnetization (M_c) in **d** a *square*, **e** a simple-cubic, **f** a *triangular* skyrmion lattice. Reprinted with permission from Ref. [16]. Copyright 2012 by American Physical Society

This also supports that a new form of skyrmion lattice, other than the triangular lattice, is realized in MnGe. However, the above calculations do not take the disarray of magnetic structure suggested by the SANS results into account, and hence the lattice form of the skyrmions in MnGe still remains inconclusive.

4.4 Summary

We observed the giant topological Hall effect as induced by the spin chirality of short-period (3 nm) skyrmion crystal emerging in MnGe. The nearly temperature-independent feature of this Hall resistivity component provides a strong evidence for the spin-chirality mechanism. In addition to the much smaller skyrmion lattice constant (3–6 nm) in MnGe than other skyrmion systems (10–200 nm), the new features revealed in this study are the tetragonal or cubic lattice form of the skyrmion crystal, not the hitherto-known hexagonal (triangular-lattice) form, and the ground-state ($H = 0$) skyrmion lattice states possibly composed of alternating skyrmions and anti-skyrmions.

References

1. M.V. Berry, Proc. R. Soc. Lond. A Math. Phys. Sci. **392**, 45 (1984)
2. N. Nagaosa, Y. Tokura, Phys. Scr. **T146**, 014020 (2012)
3. D. Xiao, M.C. Chang, Q. Niu, Rev. Mod. Phys. **82**, 1959 (2010)
4. N. Nagaosa, J. Sinova, S. Onoda, A.H. MacDonald, N.P. Ong, Rev. Mod. Phys. **82**, 1539 (2010)
5. M. Onoda, G. Tatara, N. Nagaosa, J. Phys. Soc. Jpn. **73**, 2624 (2004)
6. Y. Taguchi, Y. Oohara, H. Yoshizawa, N. Nagaosa, Y. Tokura, Science **291**, 2573 (2001)
7. Y. Machida, S. Nakatsuji, Y. Maeno, T. Tayama, T. Sakakibara, S. Onoda, Phys. Rev. Lett. **98**, 057203 (2007)
8. A. Neubauer, C. Pfleiderer, B. Binz, A. Rosch, R. Ritz, P.G. Niklowitz, P. Böni, Phys. Rev. Lett. **102**, 186602 (2009)
9. M. Lee, W. Kang, Y. Onose, Y. Tokura, N.P. Ong, Phys. Rev. Lett. **102**, 186601 (2009)
10. N. Kanazawa, Y. Onose, T. Arima, D. Okuyama, K. Ohoyama, S. Wakimoto, K. Kakurai, S. Ishiwata, Y. Tokura, Phys. Rev. Lett. **106**, 156603 (2011)
11. M. Lee, Y. Onose, Y. Tokura, N.P. Ong, Phys. Rev. B **75**, 172403 (2007)
12. B. Binz, A. Vishwanath, V. Aji, Phys. Rev. Lett. **96**, 207202 (2006)
13. U.K. Rößler, A.N. Bogdanov, C. Pfleiderer, Nature (London) **442**, 797 (2006)
14. S.D. Yi, S. Onoda, N. Nagaosa, J.H. Han, Phys. Rev. B **80**, 054416 (2009)
15. J.H. Park, J.H. Han, Phys. Rev. B **83**, 184406 (2011)
16. N. Kanazawa, J.-H. Kim, D.S. Inosov, J.S. White, N. Egetenmeyer, J.L. Gavilano, S. Ishiwata, Y. Onose, T. Arima, B. Keimer, Y. Tokura, Phys. Rev. B **86**, 134425 (2012)
17. O.L. Makarova, A.V. Tsvyashchenko, G. Andre, F. Porcher, L.N. Fomicheva, N. Rey, I. Mirebeau, Phys. Rev. B **85**, 205205 (2012)
18. M. Takeda, Y. Endoh, K. Kakurai, Y. Onose, J. Suzuki, Y. Tokura, J. Phys. Soc. Jpn. **78**, 093704 (2009)
19. S.V. Grigoriev, S.V. Maleyev, A.I. Okorokov, Y.O. Chetverikov, H. Eckerlebe, Phys. Rev. B **73**, 224440 (2006)
20. S. Okada, T. Shishido, M. Ogawa, F. Matsukawa, Y. Ishizawa, K. Nakajima, T. Fukuda, T. Lundström, J. Cryst. Growth **229**, 532 (2001)
21. X.Z. Yu, Y. Onose, N. Kanazawa, J.H. Park, J.H. Han, Y. Matsui, N. Nagaosa, Y. Tokura, Nature (London) **465**, 901 (2010)
22. U.K. Rößler, A.A. Leonov, A.N. Bogdanov, J. Phys. Conf. Ser. **303**, 012105 (2011)
23. B. Binz, A. Vishwanath, Phys. B **403**, 1336 (2008)

Chapter 5
Skyrmion Formation in Epitaxial FeGe Thin Films

5.1 Introduction

A topologically-winding spin texture of skyrmion offers characteristic emergent fields manifesting themselves in transport phenomena as detailed in Chap. 4. Spins in a skyrmion point in all directions wrapping a sphere, which is described by an integer winding number of -1 in terms of topology. Owing to the topological nature of skyrmion, emergent magnetic flux, which conduction electrons feel in a skyrmion, has a quantized value of $-\phi_0 = -h/e$, and it induces topological Hall effect (THE) besides normal and anomalous Hall effect (NHE and AHE) in skyrmionic magnets [1–6]. This emergent magnetic field enhances the appeal of skyrmions as information carriers. The skyrmion is now widely recognized as a potential information bit in next-generation magnetic non-volatile memories [7, 8].

The present hard-disk technology utilizes magnetization fragments as information bits, which shows non-volatility. The access speed is, however, intrinsically slow in the hard disk drive because the information is written and/or read out by rotating the disk. Towards the realization of the magnetic storage device with higher performance, recent studies focus on the control of magnetic structures by electric current [9]. To improve information density and high access speed, a magnetic structure with small size and quick response to electric stimulation is therefore required.

The skyrmion provides good functions for realizing such new devices owing to its topological nature: the small-sized particle nature (3–200 nm) promises ultra-high density recording [10, 11], the current-induced skyrmion motion with high mobility enables electrically-controllable information manipulation with low-energy consumption [12, 13], and creation/annhilation of skyrmions by spin polarized current demonstrates a method of writing/deleting information [14]. In the course of developing skyrmion-based memory, there is a real need for simple readout techniques of identifying individual skyrmions or counting the number of skyrmions. The emergent flux quantum inherent in each skyrmion may serve a function in detecting single skyrmions. Most notably, effective magnetic field observed in Hall voltage

N. Kanazawa, *Charge and Heat Transport Phenomena in Electronic and Spin Structures in B20-type Compounds*, Springer Theses,
DOI 10.1007/978-4-431-55660-2_5

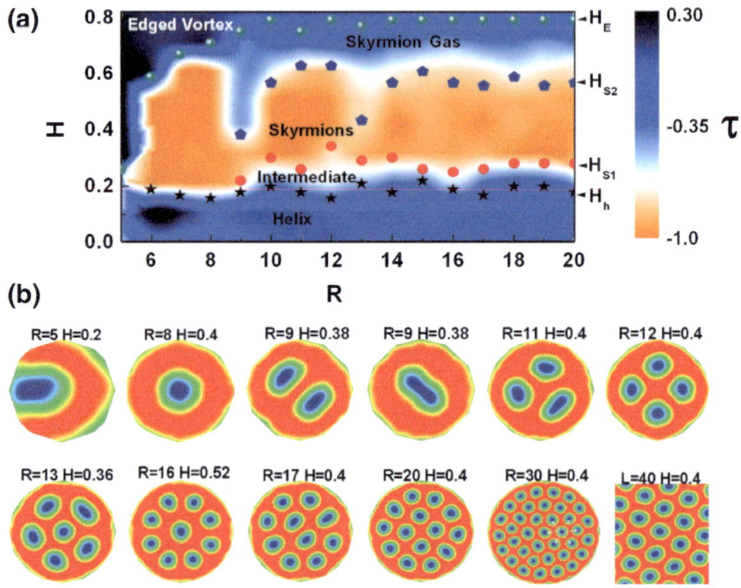

Fig. 5.1 **a** Phase diagram for development of magnetic structure in helimagnetic nanodisks as fuctions of magnetic field (H) and disk radius R. Normalized magnetic chirality $\tau = a \sum_r \hat{\mathbf{S}}_\mathbf{r} \cdot (\hat{\mathbf{S}}_{\mathbf{r}+\hat{x}} \times \hat{\mathbf{S}}_{\mathbf{r}+\hat{y}}) + \hat{\mathbf{S}}_\mathbf{r} \cdot (\hat{\mathbf{S}}_{\mathbf{r}-\hat{x}} \times \hat{\mathbf{S}}_{\mathbf{r}-\hat{y}})$ is plotted in the contour map. Skyrmion formation is observed in nanodisks with their radius $R > 5$. **b** Skyrmion formation in nanodisks of various sizes. Reprinted with permission from Ref. [18]. Copyright 2013 by American Physical Society

increases inversely with the square of skyrmion size a_{sk}, following the relation of $\mathbf{B}_{\mathrm{eff}} = -\phi_0/a_{\mathrm{sk}}^2$, which ideally reaches approximately 4000 T in a 1-nm skyrmion. This means that detection of skyrmions becomes easier when we improve areal density of information by developing smaller-sized skyrmions.

Although the quantization nature is averaged out in topological Hall signals in many-skyrmion systems of large sample sizes, which are used at previous Hall measurements in bulks and films [1–6, 15–17], it would become prominent when the system shrinks down to the size of several skyrmions. However, there is no experimental demonstration despite many theoretical predictions [18–22].[1] Here we introduce in Fig. 5.1 sample-size dependence of skyrmion formation reported by Du et al. [18]. There is a critical size (R_{c}) which determines whether skyrmion can be formed in a nanostructure. Small nanodevices with their sizes below R_{c} cannot even contain a single skyrmion. This result indicates the limitation of miniaturization of the future skyrmion devices.

[1]There are experimental reports studying finite-size effect on skyrmion formation by real-space observation [23] and resistivity measurements [24] in MnSi nanowires. However, the wire width is still too large to feature the individual skyrmions.

In this chapter, we experimentally study the skyrmion formation in variously-sized Hall-bar nanocircuits made of a FeGe thin film. As is the case of previous Hall measurement on FeGe thin films [15], topological Hall signals due to skyrmion formation are observed in the devices with circuit line width above 50 nm. Further miniaturization to 32-nm-wide circuit removes the typical signature of skyrmion formation, which indicates that skyrmions are not able to form in a smaller area than half the size of themselves. In addition, the quantized nature of the emergent field inherent in each skyrmion appears as discretized magnetic-field dependence of topological Hall resistivity in the nanoscale devices as expected.

5.2 Fabrication of FeGe Thin Films

Depending on the reaction conditions, four kinds of epitaxial films of different compounds appeared. From reflection high energy electron diffraction (RHEED) and 2θ-θ x-ray diffraction (XRD) patterns, we distinguished the respective compounds and classified the reaction conditions into four regions. Figure 5.2 shows typical XRD and RHEED patterns measured at the four thin films of different compounds. In the low temperature range (<200 °C), B20-type FeGe is not grown even if changing the deposition ratio between Fe and Ge; instead Fe_5Ge_3 and $FeGe_2$ are crystallized in the Fe-rich and Ge-rich deposition conditions, respectively. In the intermediate temperature range (300–350 °C), B20-type FeGe of purpose is epitaxially grown showing a sharp $\sqrt{3} \times \sqrt{3}$ RHEED pattern. In the high temperature range (>450 °C), Fe atoms react readily with Si substrate and form into B20-type FeSi in addition to B20-type FeGe. The extra Ge atoms appear as pure germanium crystal. RHEED pattern of the film still show $\sqrt{3} \times \sqrt{3}$ pattern, however intensity and width of the streaks are weak and blurred, which represents bad crystal quality and rough film surface.

We summarize the deposition conditions and the consequent thin films in Fig. 5.3. The above four types of thin films can be well categorized into four separated growth conditions in the plane of the growth temperature T_{growth} and Ge deposition ratio. The single phase nature of B20-type FeGe can be achieved at the narrow temperature region ($T_{\text{growth}} \approx$ 300–350 °C). 40-nm-thick FeGe thin film, which we used for the following measurements, was grown at 325 °C.

Hall-bar devices of various circuit-line-widths (32–250 nm and 10 μm) were fabricated from 40-nm-thick FeGe epitaxial thin film by using electron-beam lithography method. A top-view image of a 50-nm-wide device is shown as an example in Fig. 5.6a. The electrical leads for current and voltage were made to be same width (w); skyrmion formations in the section areas of $w \times w$ were measured by topological Hall effect.

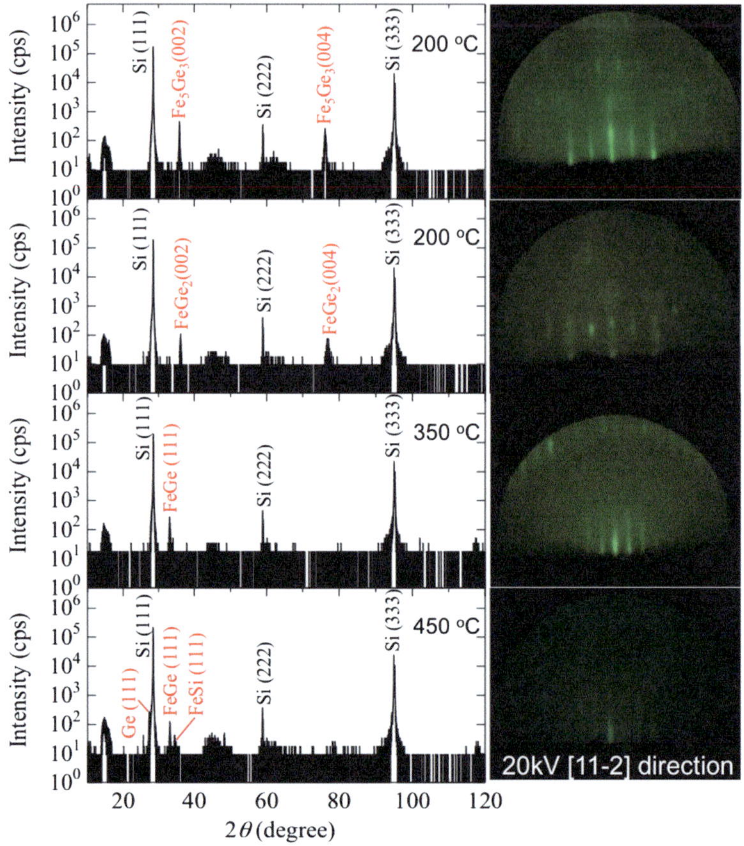

Fig. 5.2 XRD and RHEED patterns of FeGe films grown under various conditions. The single phase nature of B20-type FeGe can be achieved at the growth temperature $T_{\mathrm{growth}} \approx 300$–$350\,^{\circ}\mathrm{C}$

5.3 Topological Hall Effect and Skyrmion Formation in FeGe Nanodevices

Topological Hall signals become more distinguishable with decreasing temperature because a large contribution from anomalous Hall effect ($\propto \rho_{xx}^2$) is suppressed at low temperatures. Hereafter, we focus on results at $T = 2$ K to highlight the topological Hall contributions.

We show in Fig. 5.4 magnetization curves at 2 K of the FeGe thin film used in the present study. Magnetic-field dependence of magnetization shows a similar behavior to the B20-type magnets [25], from which we can roughly estimate the field profile of anomalous Hall resistivity. Little hysteresis loop is observed in the M–H curves at low fields, which may be associated with the switching between the skyrmion states with core-magnetization pointing up and down if it is intrinsic [15, 22].

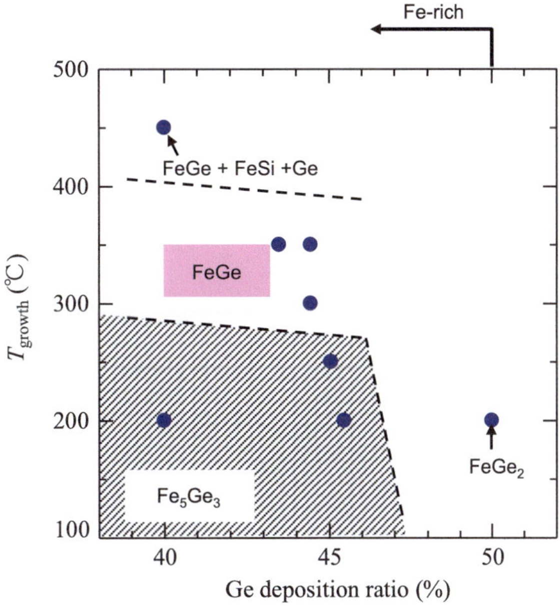

Fig. 5.3 Reaction condition map for intermetallic Fe-Ge thin films

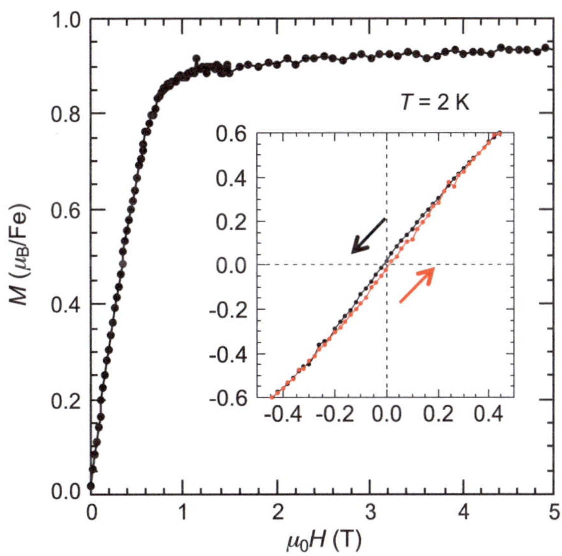

Fig. 5.4 Magnetic-field dependence of magnetization at $T = 2$ K. The inset is magnified image of magnetization curves at low fields

We show in Fig. 5.5 temperature dependence of longitudinal resistivity ρ_{xx} of the Hall-bar devices. Resistivity becomes larger in the smaller device because conduction electrons are scattered at the boundaries more frequently. A hump structure around 240 K in the resistivity of the 50-nm device is just a measurement artifact.

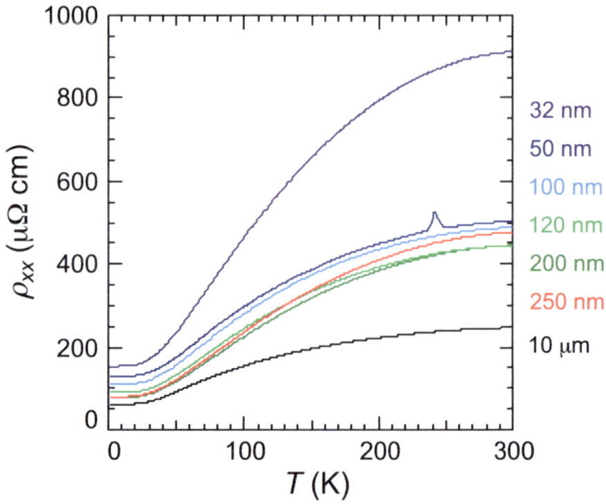

Fig. 5.5 Temperature dependence of longitudinal resistivity ρ_{xx} of the Hall-bar devices

We first confirm skyrmion formation in a 10-μm-wide FeGe device by topological Hall effect (THE). The formation of helical and/or skyrmion structures would not be influenced by the presence of the boundary condition in the 10-μm-wide device because the length scales of magnetic structure (≈70 nm) and device size are different by orders of magnitude. Topological Hall resistivity ρ_{yx}^{T} associated with formation of a non-coplanar magnetic structure is usually defined as an additional contribution other than the linearly-varying terms with magnetic field H and magnetization M, which are generally identical to normal and anomalous Hall resistivity ρ_{yx}^{N} and ρ_{yx}^{A}, respectively. However, those two components (ρ_{yx}^{N} and ρ_{yx}^{A}) can show H- and M-nonlinear dependence especially in multiband metals [17, 27].

Here we adopt a method for evaluating THE based on H-direction-sensitive skyrmion stability in thin films [17]. According to neutron scattering experiments on skyrmionic materials [28], the plane of skyrmion lattice, where the three magnetic modulation vectors **Q** lie, is strictly perpendicular to the magnetic field. Namely, energy gain from interactions among the spins on the same plane perpendicular to H is crucial for stabilization of skyrmions. When H is tilted from the normal vector of the thin film plane **n**, **Q**s have out-of-plane component, which means that spins near the surfaces lose their partner spins to interact with. The skyrmion state is thus easily destabilized with H tilted from **n** in thin films. Indeed, rapid reduction of THE above a critical inclination angle ($\theta \approx 30°$) is observed in a $Mn_{1-x}Fe_xSi$ thin film as expected [17].

Figure 5.6c shows magnetic-field dependence of Hall resistivity at various inclination angles θ at 2 K in a 10-μm-wide Hall-bar device. We observe a loop in Hall resistivity at $\theta = 0°$, which is recognized to be induced by residual skyrmions of metastable state dependent on magnetic-field history as reported by Huang and Chien [15]. The hysteresis loop drastically shrinks with a slight change in the inclination angle: the loop size becomes less than the half just tilting at $\theta = 4°$ and nearly vanishes above $\theta = 10°$. This sudden reduction of the ρ_{yx} loop, which is

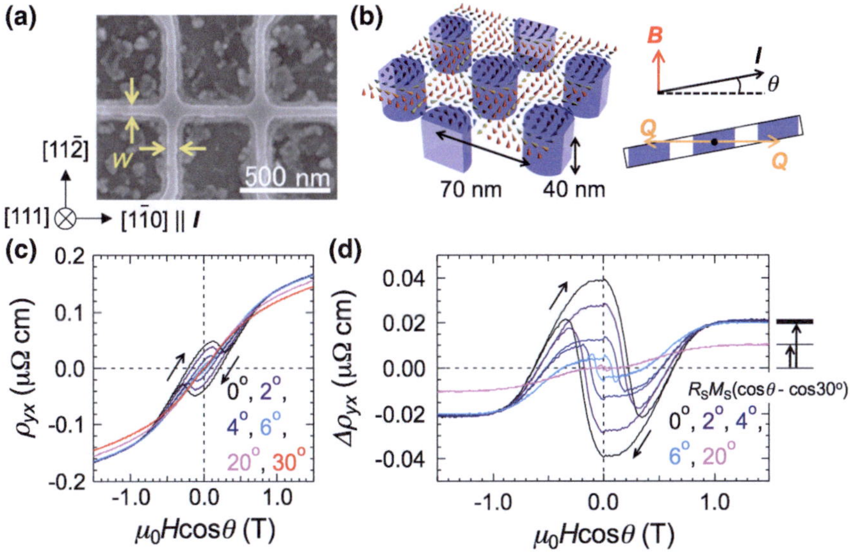

Fig. 5.6 **a** Scanning electron microscope image of a Hall-effect measurement circuit with line width of $w = 50$ nm. **b** Schematic illustrations of the skyrmion-lattice state in a 40-nm-thick FeGe thin film. The magnetic modulation period in the FeGe thin film is assumed to be the same length of 70 nm as the bulk sample. The skyrmion lattice forms with its modulation vectors **Q**s strictly perpendicular to the field **H** regardless of the field direction [28]. The skyrmion state receives less energy gain and becomes unstable when **Q**s have out-of-plane components, i.e. when the normal vector to the thin film is tilted from the magnetic field. Magnetic-field dependence of **c** Hall resistivity ρ_{yx} and **d** $\Delta\rho_{yx}$ at various inclination angles at 2 K in a 10-μm-wide circuit. (See text for the definition of $\Delta\rho_{yx}$.) A sudden shrinkage of hysteresis loops in panels (**c**) and (**d**) represents declines in skyrmion numbers. Reprinted with permission from Ref. [26]. Copyright 2015 by American Physical Society

highly unlikely in ordinary ferromagnets, would represent the destruction of two-dimensional skyrmion structures in similar to the case of $Mn_{1-x}Fe_xSi$ thin film [17]. Here we note that skyrmion stability against magnetic field direction become more significant in our FeGe thin film than that of $Mn_{1-x}Fe_xSi$. This is because the stability against inclination angle is also determined by film thickness relative to skyrmion size.

We extract topological Hall component in Fig. 5.6d by subtracting ρ_{yx} at a high angle of $\theta = 30°$ consisting of normal and anomalous Hall effect from low-angle data. If the vertical component of H to the film plane ($H_\perp$) is lower than the critical field H_c, where the ferromagnetic state is induced, the difference between Hall resistivities at low and high ($\theta = 30°$) angles is written as

$$\begin{aligned}\Delta\rho_{yx} &= \rho_{yx}(\theta) - \rho_{yx}(30^\circ)\\ &= [\rho_{yx}^{\mathrm{N}}(\theta) + \rho_{yx}^{\mathrm{A}}(\theta) + \rho_{yx}^{\mathrm{T}}(\theta)] - [\rho_{yx}^{\mathrm{N}}(30^\circ) + \rho_{yx}^{\mathrm{A}}(30^\circ)]\\ &= (R_0 B_\perp + R_{\mathrm{s}} M_\perp + R_0 \mathbf{B}_{\mathrm{eff}} \cos\theta) - (R_0 B_\perp + R_{\mathrm{s}} M_\perp)\\ &= R_0 \mathbf{B}_{\mathrm{eff}} \cos\theta.\end{aligned}$$

Otherwise ($H_\perp > H_{\mathrm{c}}$), $\Delta\rho_{yx}$ is

$$\begin{aligned}\Delta\rho_{yx} &= [\rho_{yx}^{\mathrm{N}}(\theta) + \rho_{yx}^{\mathrm{A}}(\theta)] - [\rho_{yx}^{\mathrm{N}}(30^\circ) + \rho_{yx}^{\mathrm{A}}(30^\circ)]\\ &= (R_0 B_\perp + R_{\mathrm{S}} M_{\mathrm{s}} \cos\theta) - (R_0 B_\perp + R_{\mathrm{S}} M_{\mathrm{s}} \cos 30^\circ)\\ &= R_{\mathrm{S}} M_{\mathrm{s}} (\cos\theta - \cos 30^\circ).\end{aligned}$$

Here R_0 and R_{S} are normal and anomalous Hall coefficient, $M_\perp$ and M_{s} are the vertical component of magnetization M and the saturated magnetization, respectively. Although the extracted $\Delta\rho_{yx}$ inevitably includes anomalous Hall component $R_{\mathrm{S}} M_{\mathrm{s}}(\cos\theta - \cos 30^\circ)$ at high fields, $\Delta\rho_{yx}$ is almost equivalent to ρ_{yx}^{T}. The rapid shrinkage of ρ_{yx}^{T} is again highlighted in Fig. 5.6d. As magnetization at the core of the skyrmion is anti-parallel to the field [28, 29], skyrmions with positive (negative) core-magnetization is stabilized at negative (positive) magnetic field. Hereafter, we call skyrmions with positive (negative) core-magnetization as core-up (down) skyrmions, which produce positive (negative) emergent magnetic fields. The hysteresis loop of ρ_{yx}^{T} demonstrates that core-up (down) skyrmions persists at zero magnetic field in the process of increasing (decreasing) field from the large negative (positive) field.

We show the line-width dependence of Hall resistivity in Fig. 5.7. While devices with wider circuit lines than 50 nm show the characteristic hysteresis loops representing skyrmion formation, Hall resistivity in a 32-nm-wide circuit almost traces the magnetization curve (see also Fig. 5.4). The collapse of the hysteresis loop with narrowing the circuit line represents that the threshold size for the formation of skyrmions lies between 32 nm and 50 nm in FeGe epitaxially grown on a Si(111) substrate.[2] As the helical period in the bulk FeGe is 70 nm, this result indicates the limit of the miniaturization of the possible skyrmion devices is around half the size of each skyrmion.

Getting a close look at the hysteresis loops in the nano-scale circuits, Hall resistivity shows discontinuous changes with field variation, which indicates the discrete change in emergent magnetic field.

Figure 5.8 shows magnified drawings of hysteresis loops and difference between Hall resistivities in increasing and decreasing field processes in various-sized devices. The latter of them is denoted by $\Delta\rho_{yx}^{\mathrm{T}}$ and plotted as normalized data by its maximum value in the figure. The conventional Hall components, i.e., ρ_{yx}^{N} and ρ_{yx}^{A},

[2]Unavoidable side damage in the etching process would have greater influence on the smaller devices, which may be a crucial damage in the 32-nm device related to the inhibition of skyrmion formation. However, considering the magnetization property of FeGe remains in the Hall resistivity in the 32-nm device, there should be some effects of miniaturization on suppression of skyrmion formation, although we could not estimate the exact value of the threshold size.

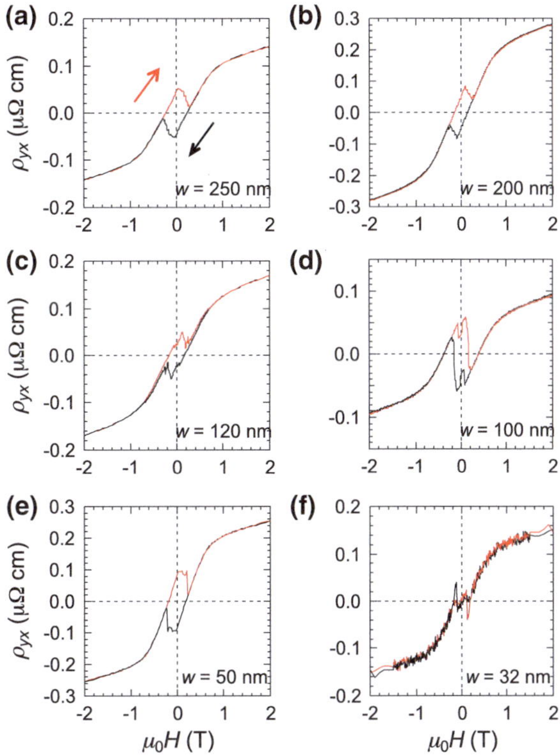

Fig. 5.7 a–f Magnetic-field dependence of Hall resistivity of FeGe nanoscale circuits of various widths ranging from 250 to 32 nm. We subtracted the H-symmetric component using the equation $\rho_{yx}(\pm H) = \pm\frac{[V_y(H)-V_y(-H)]t}{2I}$. Reprinted with permission from Ref. [26]. Copyright 2015 by American Physical Society

remain unchanged against magnetic field history because there are no hystereses in the carrier density and magnetization (Fig. 5.4). The hysteretic Hall signal $\Delta\rho_{yx}^{\mathrm{T}}$ represents an additional contribution other than NHE and AHE, which we attribute to the topological Hall signal. In contrast to the smooth variation of $\Delta\rho_{yx}^{\mathrm{T}}$ in the 10-μm device (Fig. 5.8a), $\Delta\rho_{yx}^{\mathrm{T}}$ in nanocircuits of $w = 50$–250 nm exhibit stepwise profiles (Fig. 5.8b–f). Constricted geometry comparable to the size of a few skyrmions emphasize their individuality, and thus their behaviors, such as deformation, creation/annihilation, and movement, become pronounced appearing as discrete signals in Hall effect via emergent magnetic field, while those are averaged to be indistinguishable in large samples containing many skyrmions.

The step heights display considerable variation and are not equal to one or integer multiple units of topological Hall resistivity $\rho_{yx}^{\mathrm{T}} = -R_0\phi_0/w^2$, which indicates additional contributions other than the discrete change in emergent magnetic field associated with creation/annihilation of skyrmions. This dispersion of topological Hall steps may originate from the discontinuous motion of skyrmions due to trapping by or releasing from impurity or defect sites in the course of field changes, as is the case with discontinuous change in magnetization observed ferromagnetic domains, so-called Barkhausen effect. The possible fluctuation of skyrmions' positions, especially

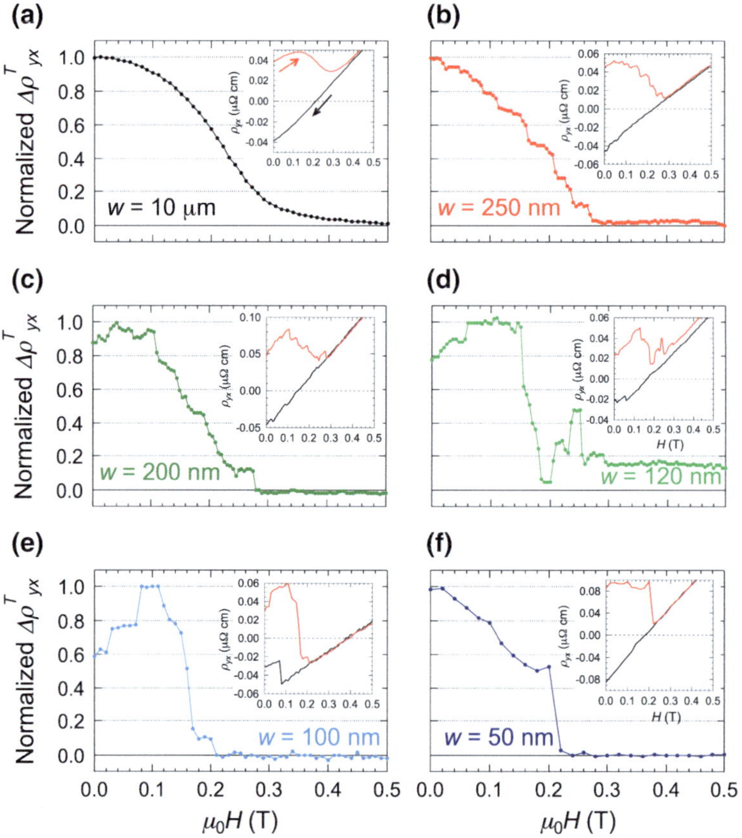

Fig. 5.8 **a–f** Magnetic-field dependence of difference between Hall resistivities in increasing and decreasing field processes, which is denoted by $\Delta\rho^T_{yx}$. Normalized $\Delta\rho^T_{yx}$ by its maximum value is presented. Insets are magnified images of each ρ_{yx} at low fields. Reprinted with permission from Ref. [26]. Copyright 2015 by American Physical Society

skyrmions putting their feet on and off the verge of the probed square area of the Hall device, may produce "halfway" discrete topological Hall effect. In particular, we attribute the fluctuation of ρ^{T}_{yx} around $\mu_0 H = 0.2$–0.3 T in 120-nm-wide device (Fig. 5.8d) and dip structure of ρ^{T}_{yx} around $\mu_0 H = 0$–0.1 T in 100-nm-wide device (Fig. 5.8e) to discontinuous changes in skyrmion position. Those fluttering magnetic-field dependence of topological Hall signal are reproduced in different magnetic field scans, indicating reversibility of magnetization process. To observe "quantized" THE, i.e., the digital changes in THE, in a skyrmion system, Hall measurements on isolated skyrmion nanodots connected with non-magnetic electrical leads for current and voltage paths are required.

We speculate skyrmion formation process in the nanostructured Hall bars based on Hall resistivity data. Here we take, for example, the case of 250-nm-wide device,

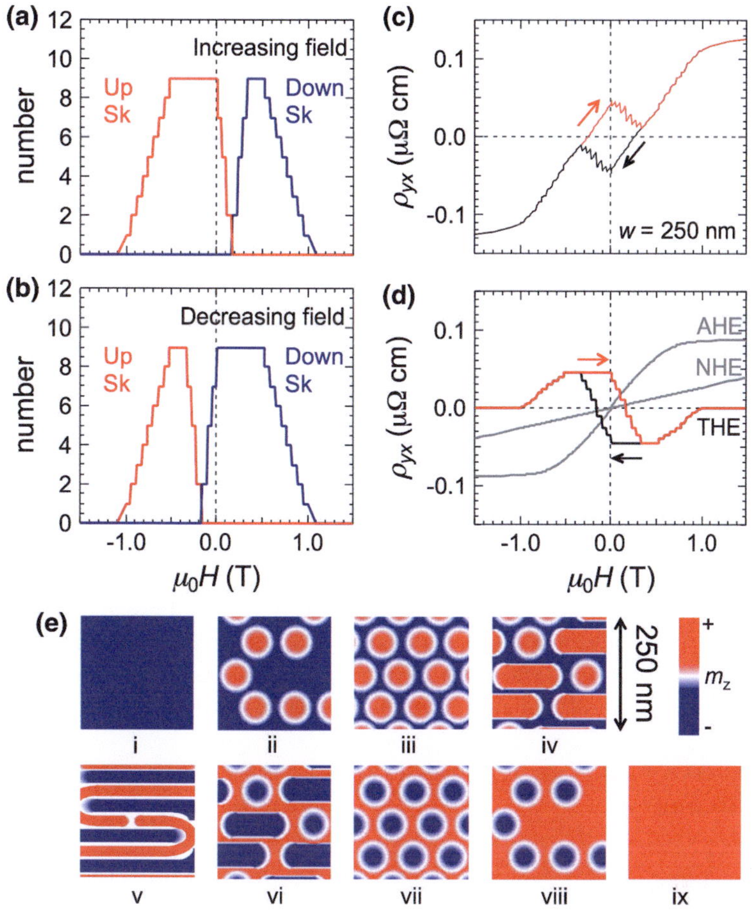

Fig. 5.9 Conceptual diagrams for skyrmion formation and corresponding Hall resistivities in a 250-nm-wide nano-scale circuit. Variations of numbers of core-up and core-down skyrmions as functions of magnetic field in **a** increasing and **b** decreasing field processes. Corresponding **c** Hall resistivity and **d** its decomposition into ρ_{yx}^{N} (NHE), ρ_{yx}^{A} (AHE), and ρ_{yx}^{T} (THE). **e** Series of schematic illustrations representing an expected development of magnetic structure as a function of magnetic field. Reprinted with permission from Ref. [26]. Copyright 2015 by American Physical Society

which can accommodate approximately 9 skyrmions at the overlap of current and voltage leads $w \times w$. Figure 5.9a, b are the conceivable development of numbers of core-up and -down skyrmions in increasing and decreasing field processes, respectively. In the increasing magnetic field from large negative field ($<H_c$) (Fig. 5.9a), the magnetic state traces stages illustrated in Fig. 5.9e as the following sequence: core-up skyrmions gradually pop up (panel ii of Fig. 5.9e) in the background of ferromagnetic state of down-pointed moments (i), high-density core-up skyrmions, possibly skyrmion lattice state, is realized and remains around zero field as the metastable state (iii), core-up skyrmions are destabilized to rapidly disappear coalescing with

neighboring ones (iv), coalesced skyrmions forms a convoluted helical structure (v), the helical structure fragments to form core-down skyrmions (vi), which then crystallize (vii) and begin to disappear (viii), and finally spins are fully polarized to be ferromagnetic state (ix). The above expectation is a reasonable model judging from the magnetic-field dependence of skyrmion formation in thinned plates of B20-type magnets revealed by the real-space observation using Lorentz transmission electron microscopy (see also Chap. 1) [29].

Magnetic-field profile of Hall resistivity is directly estimated from the assumed skyrmion formation as presented in Fig. 5.9c, d. Here we do not take into account the discontinuous motion of skyrmions. A unit of topological Hall resistivity produced by a single skyrmion is determined from $\Delta\rho_{yx}^{\mathrm{T}}$: As we now assume the maximum value of $\Delta\rho_{yx}^{\mathrm{T}}$ represent the difference between topological Hall resistivities produced by close-packed core-up and -down skyrmions, the unit of ρ_{yx}^{T} is $\Delta\rho_{yx}^{\mathrm{T}}$ divided by twice the maximum capacity of the overlap area ($w \times w$). In the present example, the unit of ρ_{yx}^{T} is approximately 5 nΩ cm of $\Delta\rho_{yx}^{\mathrm{T}}/(2 \times 9)$. We present corresponding ρ_{yx}^{T} to the speculated skyrmion formation in Fig. 5.9d along with NHE and AHE evaluated from the slope of Hall resistivity at high fields above H_{c} and magnetization M, respectively. The overall profile of ρ_{yx}^{T} bears a remarkable resemblance to the extracted ρ_{yx}^{T} of the 10-μm device, which supports our model. The total Hall resistivity (Fig. 5.9c) also shows a good agreement with the experimental results (Fig. 5.7). The discretized feature of ρ_{yx}^{T} especially shows up in parts of the hysteresis loop where the rapid switch between core-up and -down skyrmions occurs, which we focus on in Fig. 5.8.

5.4 Summary

We demonstrated the skyrmion formation in nanostructured FeGe Hall-bar devices with their sizes of one or a few skyrmions by measurements of topological Hall signals. In particular, the discrete changes in topological Hall resistivity represent the quantized nature of the emergent magnetic field inherent in each skyrmion, which gets buried in many other skyrmions residing in large-sized samples used in prior experimental studies. These results show that the characteristic emergent field will serve as a detection method of single skyrmions in the future skyrmion-based memories consisting of nanocircuits.

References

1. A. Neubauer, C. Pfleiderer, B. Binz, A. Rosch, R. Ritz, P.G. Niklowitz, P. Böni, Phys. Rev. Lett. **102**, 186602 (2009)
2. M. Lee, W. Kang, Y. Onose, Y. Tokura, N.P. Ong, Phys. Rev. Lett. **102**, 186601 (2009)
3. N. Kanazawa, Y. Onose, T. Arima, D. Okuyama, K. Ohoyama, S. Wakimoto, K. Kakurai, S. Ishiwata, Y. Tokura, Phys. Rev. Lett. **106**, 156603 (2011)
4. B.J. Chapman, M.G. Grossnickle, T. Wolf, M. Lee, Phys. Rev. B **88**, 214406 (2013)

5. R. Ritz, M. Halder, M. Wagner, C. Franz, A. Bauer, C. Pfleiderer, Nature (London) **497**, 231 (2013)
6. C. Franz, F. Freimuth, A. Bauer, R. Ritz, C. Schnarr, C. Duvinage, T. Adams, S. Blügel, A. Rosch, Y. Mokrousov, C. Pfleiderer, Phys. Rev. Lett. **102**, 186601 (2014)
7. A. Fert, V. Cros, J. Sampaio, Nat. Nanotech. **8**, 152 (2013)
8. N. Nagaosa, Y. Tokura, Nat. Nanotech. **8**, 899 (2013)
9. S.S.P. Parkin, M. Hayashi, L. Thomas, Science **320**, 190 (2008)
10. S. Heinze, K. von Bergmann, M. Menzel, J. Brede, A. Kubetzka, R. Wiesendanger, G. Bihlmayer, S. Blügel, Nat. Phys. **7**, 713 (2011)
11. K. Shibata, X.Z. Yu, T. Hara, D. Morikawa, N. Kanazawa, K. Kimoto, S. Ishiwata, Y. Matsui, Y. Tokura, Nat. Nanotech. **8**, 723 (2013)
12. F. Jonietz, S. Mühlbauer, C. Pfleiderer, A. Neubauer, W. Münzer, A. Bauer, T. Adams, R. Georgii, P. Böni, R.A. Duine, K. Everschor, M. Garst, A. Rosch, Science **330**, 1648 (2010)
13. X.Z. Yu, N. Kanazawa, W.Z. Zhang, T. Nagai, T. Hara, K. Kimoto, Y. Matsui, Y. Onose, Y. Tokura, Nat. Commun. **3**, 988 (2012)
14. N. Romming, C. Hanneken, M. Menzel, J.E. Bickel, B. Wolter, K. von Bergmann, A. Kubetzka, R. Wiesendanger, Science **341**, 636 (2013)
15. S.X. Huang, C.L. Chien, Phys. Rev. Lett. **108**, 267201 (2012)
16. Y. Li, N. Kanazawa, X.Z. Yu, A. Tsukazaki, M. Kawasaki, M. Ichikawa, X.F. Jin, F. Kagawa, Y. Tokura, Phys. Rev. Lett. **110**, 117202 (2013)
17. T. Yokouchi, N. Kanazawa, A. Tsukazaki, Y. Kozuka, M. Kawasaki, M. Ichikawa, F. Kagawa, Y. Tokura, Phys. Rev. B **89**, 064416 (2014)
18. H. Du, W. Ning, M. Tian, Y. Zhang, Phys. Rev. B **87**, 014401 (2013)
19. L. Sun, R.X. Cao, B.F. Miao, Z. Feng, B. You, D. Wu, W. Zhang, An Hu, H.F. Ding, Phys. Rev. Lett. **110**, 167201 (2013)
20. J. Sampaio, V. Cros, S. Rohart, A. Thiaville, A. Fert, Nat. Nanotech. **8**, 839 (2013)
21. S. Rohart, A. Thiaville, Phys. Rev. B **88**, 184422 (2013)
22. M. Beg, D. Chernyshenko, M.A. Bisotti, W.W. Wang, M. Albert, R.L. Stamps, H. Fangohr, arXiv: 1312.7665v2
23. X.Z. Yu, J.P. DeGrave, Y. Hara, S. Jin, Y. Tokura, Nano Lett. **13**, 3755 (2013)
24. H. Du, J.P. DeGrave, F. Xue, D. Liang, W. Ning, J. Yang, M. Tian, Y. Zhang, S. Jin, Nano Lett. **14**, 2026 (2014)
25. Y. Onose, N. Takeshita, C. Terakura, H. Takagi, Y. Tokura, Phys. Rev. B **72**, 224431 (2005)
26. N. Kanazawa, M. Kubota, A. Tsukazaki, Y. Kozuka, K.S. Takahashi, M. Kawasaki, M. Ichikawa, F. Kagawa, Y. Tokura, Phys. Rev. B **91**, 041122(R) (2015)
27. Y. Taguchi, Y. Oohara, H. Yoshizawa, N. Nagaosa, Y. Tokura, Science **291**, 2573 (2001)
28. S. Mühlbauer, B. Binz, F. Jonietz, C. Pfleiderer, A. Rosch, A. Neubauer, R. Georgii, P. Böni, Science **323**, 915 (2009)
29. X.Z. Yu, Y. Onose, N. Kanazawa, J.H. Park, J.H. Han, Y. Matsui, N. Nagaosa, Y. Tokura, Nature (London) **465**, 901 (2010)

Chapter 6
3D Dirac Electrons and Large Thermoelectric Properties in CoGe

6.1 Introduction

6.1.1 *Thermoelectric Figure of Merit*

A thermoelectric material, which can convert waste heat to electricity or pump heat as a Peltier cooler, is a candidate for a sustainable energy solution and some materials with high efficiency have been already applied for our life. Developing highly efficient thermoelectrics is one of the most demanding targets in materials science. The efficiency of a thermoelectric is measured by the figure of merit $ZT = S^2T/(\rho\kappa)$, where S is Seebeck coefficient, ρ is resistivity, κ is thermal conductivity, and T is temperature. Larger ZT corresponds to higher efficiency, i.e., a material with a large Seebeck coefficient, a low resistivity, and a low thermal conductivity is desired. A rough criterion for possible applications is $ZT \geq 1$.

Heavily-doped multi-valley semiconductors have been extensively studied according to the well-known guiding principle that the nearly-degenerate carrier concentration of $\approx 10^{19}\,\mathrm{cm}^{-3}$ maximizes the figure of merit $ZT = S^2T/(\rho\kappa)$ in the single parabolic band model (Fig. 6.1) [1]. Almost all the metals are not attractive thermoelectrics for their low Seebeck coefficient (S) on the order of a few $\mu\mathrm{V/K}$.

6.1.2 *Boltzmann Transport Theory in Metals*

There are, however, some metallic materials with one order of magnitude larger thermopower than ordinary metals, for example, Heusler compounds (e.g. Fe_2VAl [2]), rare-earth compounds (e.g. $CePd_3$ [3] and $YbAl_3$ [4]), and layered cobalt oxides (e.g. Na_xCoO_2 [5]). Here, to identify the origin of the large Seebeck coefficients, we briefly review the Boltzmann transport theory in metals [6].

N. Kanazawa, *Charge and Heat Transport Phenomena in Electronic and Spin Structures in B20-type Compounds*, Springer Theses,
DOI 10.1007/978-4-431-55660-2_6

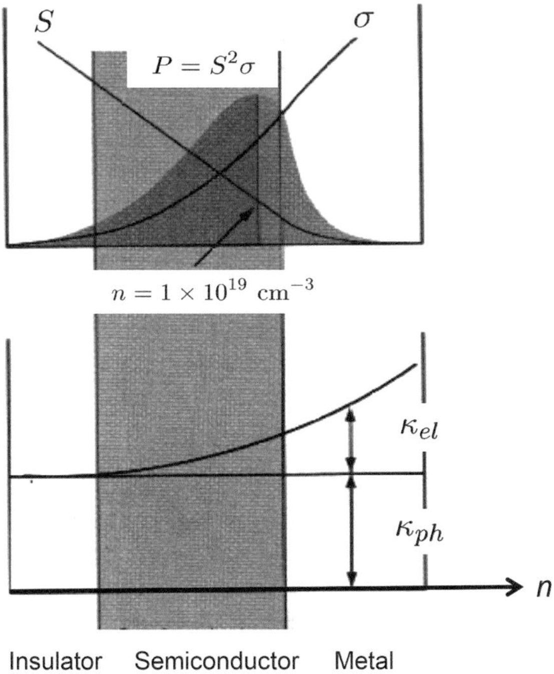

Fig. 6.1 Carrier density dependence of Seebeck coefficient, conductivity, and thermal conductivity in the system with a parabolic band structure

The Boltzmann transport equation is described as

$$\frac{\partial g_n}{\partial t} + \mathbf{v}_n \cdot \frac{\partial g_n}{\partial \mathbf{r}} + \frac{\mathbf{F}}{\hbar} \cdot \frac{\partial g_n}{\partial \mathbf{k}} = \left. \frac{\partial g_n}{\partial t} \right|_{coll}, \tag{6.1}$$

where g_n and g_0 are the electron distribution functions at time t and for the equilibrium state, respectively, and n is the band index.

In the semiclassical approximation, we adopt the following assumptions: (i) ignoring inter-band transitions. (ii) the position $\mathbf{r}$ and the wave vector $\mathbf{k}$ can be described by equations of motions $\frac{d\mathbf{r}}{dt} = \frac{1}{\hbar}\frac{\partial \varepsilon_n(\mathbf{k})}{\partial \mathbf{k}}$ and $\hbar \frac{d\mathbf{k}}{dt} = -e\left[\mathbf{E}(\mathbf{r}, t) + \mathbf{v}_n \times \mathbf{B}(\mathbf{r}, t)\right]$, respectively. (iii) the equilibrium distribution function g_0 is Fermi distribution function $g^0[\varepsilon_n(\mathbf{k})] = 1/[e^{(\varepsilon_n(\mathbf{k})-\mu)/k_B T} + 1]$. Finally, the Boltzmann equation is transformed to

$$\frac{\partial g_n}{\partial t} + \mathbf{v}_n \cdot \frac{\partial g_n}{\partial \mathbf{r}} + \frac{\mathbf{F}}{\hbar} \cdot \frac{\partial g_n}{\partial \mathbf{k}} = -\frac{g_n - g_n^0}{\tau_n(\mathbf{k})}. \tag{6.2}$$

From the Boltzmann equation, electrical conductivity σ, thermal conductivity κ, and Seebeck coefficient S are deduced as

$$\sigma = L^{11} \approx \frac{ne^2\tau}{m} \tag{6.3}$$

$$\kappa = L^{22} - L^{21}(L^{11})^{-1}L^{12} = L^{22} + O\left(\frac{k_B T}{\varepsilon_F}\right)^2 \approx \frac{\pi^2}{3}\left(\frac{k_B}{e}\right)^2 T\sigma \tag{6.4}$$

$$S = L^{12}(L^{11})^{-1} \approx \frac{\pi^2}{3}\frac{k_B^2}{e}T\left[\frac{\partial \ln\sigma(\varepsilon)}{\partial\varepsilon}\right]_{\varepsilon=E_F}. \tag{6.5}$$

In particular, Eqs. 6.4 and 6.5 are known as the Wiedemann-Franz law and the Mott relation, respectively.

For designing a large Seebeck coefficient, however, the derived Mott relation blurs the essence of the Boltzmann transport theory. An expression derived on the way to the Mott relation is more useful [7]:

$$S = -\frac{1}{eT}\frac{e^2\tau\int\frac{d\mathbf{k}}{4\pi^3}\left(-\frac{\partial g^0}{\partial\varepsilon}\right)\mathbf{v}(\mathbf{k})\mathbf{v}(\mathbf{k})[\varepsilon(\mathbf{k})-\mu]}{e^2\tau\int\frac{d\mathbf{k}}{4\pi^3}\left(-\frac{\partial g^0}{\partial\varepsilon}\right)\mathbf{v}(\mathbf{k})\mathbf{v}(\mathbf{k})}. \tag{6.6}$$

Because the energy derivative of the fermi distribution function is roughly regarded as a pulse-like function

$$-\frac{\partial f(\varepsilon)}{\partial\varepsilon} = \begin{cases} 1\ (-k_B T \le \varepsilon \le k_B T) \\ 0\ (\text{otherwise}) \end{cases}$$

we simplify the numerator of Eq. 6.6 as

$$\begin{aligned} S &\propto (\varepsilon_{\text{up}} - E_F)\tau_{\text{up}}D_{\text{up}}v_{\text{up}}^2 + (\varepsilon_{\text{down}} - E_F)\tau_{\text{down}}D_{\text{down}}v_{\text{down}}^2 \\ &\approx k_B T(\tau_{\text{up}}D_{\text{up}}v_{\text{up}}^2 - \tau_{\text{down}}D_{\text{down}}v_{\text{down}}^2). \end{aligned} \tag{6.7}$$

Here the subscripts "up" and "down" represent the electron states whose energy levels lie above and below E_F, respectively. τ, D, and v are relaxation time, density of state, and group velocity. Equation 6.7 implies that S in metal is governed by the asymmetry of the density of state $D(E)$, the relaxation time $\tau(E)$, and the group velocity $v(E)$ of the electron near the Fermi energy (E_F). In other words, it is important to avoid the cancellation of thermal energy transport between the contributions from electrons and holes around E_F by designing the strong asymmetry between them.

In fact, the above-exemplified metallic materials show asymmetric band structure [7, 8] or large energy-dependence of τ around E_F [1]. Incidentally, we note that there is another interpretation of thermoelectric profiles in Na_xCoO_2 which attributes an enhancement of carried entropy to the spin and orbital degeneracy [9].

6.1.3 Thermoelectric Properties of CoSi

Among good thermoelectric compounds, transition metal silicides, such as $MnSi_{1.7}$, $FeSi_2$, and CoSi, are well-known potential thermoelectrics, which have an advantage in resource availability and robustness for mechanical shocks and chemical reactions. CoSi with B20-type crystal structure of our focus is a metallic thermoelectrics with one order of magnitude larger thermopower than ordinary metals (about $-80\,\mu V/K$ at room temperature) [10, 11], which also originates from an asymmetric band structure with a pseudogap [12]. In addition, a p-type CoSi with large positive thermopower can be realized by partial substitution of Co with Fe [10–12], Si with Al [11, 13, 14], and so on [15], which takes an advantage for constructing a typical module for generators or refrigerators, i.e., a thermoelectric couple consisting of n-type and p-type thermoelectric elements [1]. Additional improvements for the thermoelectric efficiency in CoSi are expected by reduction of thermal conductivity: scattering phonons by alloying, creating interfaces (fabricating a superlattice), and using composites mixed on the nanometer scale. To that end, it is very natural to use another B20-type compound with the same electron number as an alloying partner, a layer in a superlattice, and a component in a nanostructure composite, respectively. Indeed, it is demonstrated that replacing Si with Ge reduces the thermal conductivity of CoSi [16, 17]. However, there is no guarantee that the asymmetric band structure remains in those materials, and this remains elusive partly because B20-type CoGe cannot be synthesized at ambient pressure [18, 19].

6.2 Thermoelectric Properties of CoGe

We have measured thermoelectric properties and calculated the band structure and thermopower in B20-type CoGe and its carrier doped compounds $Co_{1-x}Fe_xGe$ and $Co_{1-y}Ni_yGe$ [20]. CoGe is also identified as a promising thermoelectric material showing both metallic low resistivity and large thermopower as well as CoSi. Furthermore, we have demonstrated that the asymmetric band structure composed of Dirac cone and flat band dispersion, which we attribute to the origin of the large magnitude of Seebeck coefficient, still remains in CoGe from the first-principles calculation and experimental results.

6.2.1 Calculations of Band Structure and Thermopower

We performed the first-principles band-structure calculations by using the full-potential linearized augumented plane wave method implemented in WIEN2K [21]. The lattice constant $a = 4.63$ Å and the relative parameter $x_{\mathrm{Co}} = 0.140$ and $x_{\mathrm{Ge}} = 0.845$ were used. The generalized gradient approximation by Perdew et al. [22] was used to take into account the exchange-correlation energy. The muffin-tin radii R_{MT} are set to 2.29 and 2.03 a.u. for Co and Ge atoms, respectively. Wave functions

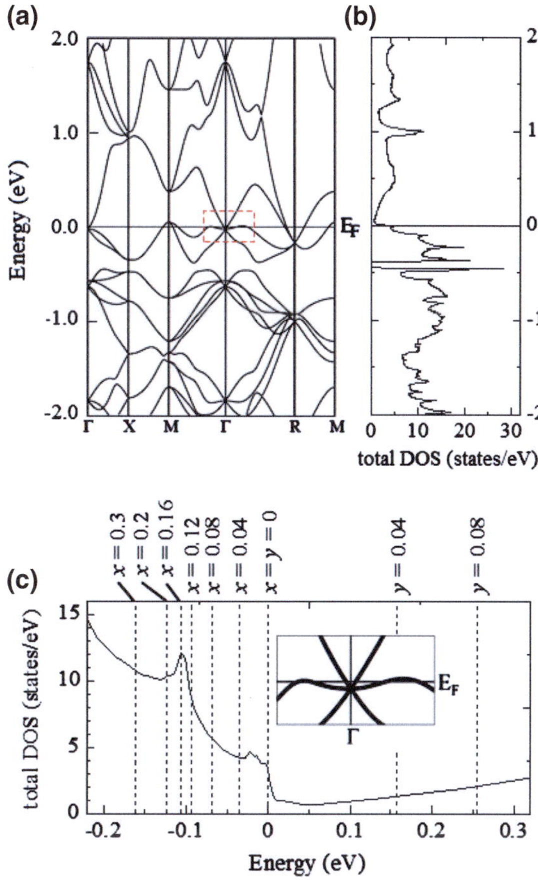

Fig. 6.2 **a** Calculated band structure, **b** density of states of the B20-type CoGe. The Fermi energy is taken at the origin (0 eV) of energy. **c** Each energy value of the Fermi level in the band-filling controlled sample, $Co_{1-x}Fe_xGe$ and $Co_{1-y}Ni_yGe$, is indicated by the *vertical dashed lines* on the basis of the rigid band approximation. The inset shows the band structure near E_F ($x = y = 0$) around the Γ point, composed of the Dirac cone and flat band with sharp bends, which corresponds to the region indicated by the *red dashed square* in panel (**a**). Reprinted with permission from Ref. [20]. Copyright 2012, American Institute of Physics

inside atomic spheres are expanded in spherical harmonics up to $l_{\max} = 10$, while those outside spheres are expanded in plane waves with the cutoff $R_{\mathrm{MT}}K_{\max} = 7.0$. The Seebeck coefficient (S) is calculated by semi-classical Boltzmann theory with the relaxation-time approximation as implemented in the BOLTZTRAP code [23].

Figure 6.2a, b show the band structure and the density of states (DOS) of CoGe, respectively. As is clear from the energy (E) dependence of DOS (Fig. 6.2b), the band structure is asymmetric around the Fermi energy, similar to that of CoSi [12]. In particular, the Dirac cone like dispersion, whose cross point locates close to E_F, is discerned around Γ point together with the much-less dispersing (flat) band which rapidly bends around E_F (see also the inset of Fig. 6.2c for magnification). Owing to this, the abrupt change of DOS exists around the Fermi energy (Fig. 6.2b), giving rise to a pseudo-gap like feature.

Dirac-cone band structure can be the origin of a large negative (positive) thermopower; this is because a small shift of E_F up above (down below) its crossing point causes an asymmetric E-dependence of DOS around E_F.

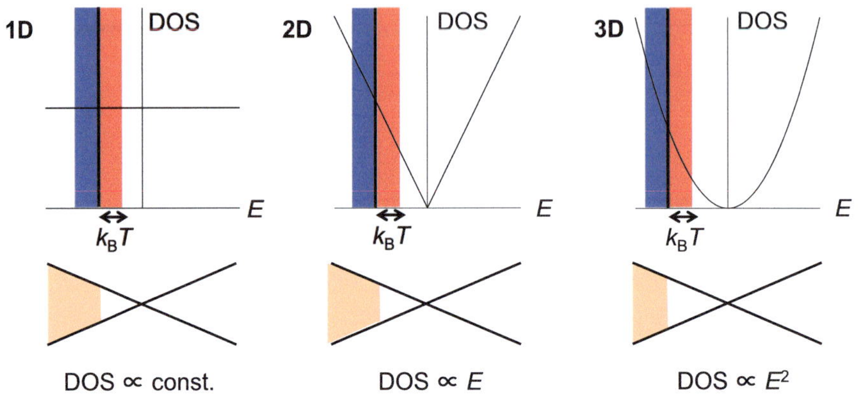

Fig. 6.3 Schematic illustrations of energy (E) dependences of DOS in 1D, 2D, and 3D Dirac electron systems, which are proportional to E^0, E^1, and E^2, respectively. For comparison, DOS in usual metallic systems (3D) with a parabolic band structure is proportional to $\sqrt{E}$

Figure 6.3 shows E-dependences of DOS in 1D, 2D, and 3D Dirac electron systems. DOS in 1D, 2D, and 3D systems depend proportional to E^0, E^1, and E^2, respectively. A shift of E_F from the Dirac point gives rise to difference between DOS above and below E_F in the 2D and 3D systems, while the constant DOS with energy variation in the 1D system cannot show any asymmetric profiles. In usual metals with parabolic band structures, DOS is proportional to $\sqrt{E}$ and E_F positions far above the valley (the bottom of the band); DOS above and below E_F is almost equal to each other. In contrast, a strong asymmetry between them is expected in 2D and 3D Dirac electron systems, which is the source of a large S.

Indeed, calculated S in a single Dirac band structure shows large positive and negative S when E_F is shifted from the crossing point. Seebeck coefficient in Dirac systems under the relaxation time approximation[1] can be described as

$$S = -\frac{1}{eT}\frac{\int\left(-\frac{\partial f(\varepsilon)}{\partial\varepsilon}\right)(\varepsilon - E_F)\varepsilon^{d-1}\mathrm{d}\varepsilon}{\int\left(-\frac{\partial f(\varepsilon)}{\partial\varepsilon}\right)\varepsilon^{d-1}\mathrm{d}\varepsilon}, \tag{6.8}$$

where d is dimension of the system. Here we note that S does not depend on the group velocity of Dirac electrons (the slope of the energy dispersion) because it is cancelled in the numerator and denominator. Figure 6.4 presents Fermi energy dependence of Seebeck coefficient in 2D and 3D Dirac cones. When E_F is at the crossing point ($E_F = 0\,\mathrm{eV}$), the cancellation between positive and negative contributions to S occurs. Large positive and negative S can be achieved if E_F is shifted from the crossing point. The extremum magnitudes of S are approximately 90 and 150 μV/K respectively in 2D and 3D system, which is a more than one order larger value than

[1]Relaxation times τ for all the electron states are assumed to be identical.

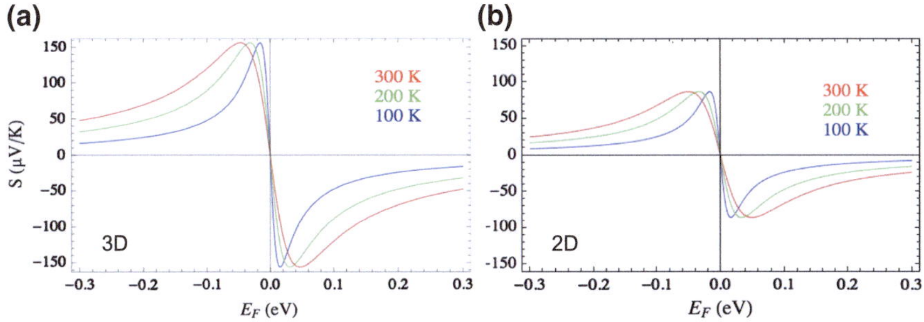

Fig. 6.4 Theoretical calculation of Fermi energy dependence of Seebeck coefficient in **a** 3D and **b** 2D Dirac materials. The origin of energy ($E_F = 0\,\mathrm{eV}$) is set to locate at the Dirac point

usual metals. In addition, in the case of CoGe, a positive contribution to S is also expected from the flat band with sharp bends due to abrupt change in group velocity around E_F.

6.2.2 Filling Dependence of Thermoelectric Properties in CoGe

Keeping the above theoretical idea in mind, here we get back to the CoGe case. The temperature profiles of resistivity ρ, thermal conductivity κ, and S of CoGe and Fe-doped ($x = 0.04$–0.3) or Ni-doped ($y = 0.04, 0.08$) samples are shown in Fig. 6.5. CoGe shows metallic resistivity, which is about $220\,\mu\Omega\,\mathrm{cm}$ at 300 K. Whereas the almost same values of resistivity are observed for the Ni-doped samples with non-doped CoGe, a sudden increase of resistivity is observed with doping Fe; this is perhaps due to the replacement of high-mobility electrons with low-mobility holes due to Fe-doping as the major charge carriers. The similar semimetallic behavior was also confirmed in Fe-doped CoSi [10, 11]. In particular, the nearly flat band around the Γ point near E_F (Fig. 6.2a) may endow the introduced holes with a large effective mass. The κ shows the typical T-dependence of the phonon thermal conductivity: At low temperatures, κ increases with increasing temperature and shows the peak around 50 K. At higher temperatures, κ decreases with increasing temperature due to the phonon-phonon scattering. Indeed, the phonon thermal conductivity dominates the total thermal conductivity; the electric contribution to the thermal conductivity is estimated less than 4 W/Km below 300 K in all the samples as deduced from the Wiedemann-Franz law $\kappa_e = L_0 \sigma T$, where σ is electric conductivity and L_0 is the Lorenz number $2.44 \times 10^{-8}\,\mathrm{W}\Omega/\mathrm{K}^2$. CoGe shows a large negative S, $-82\,\mu\mathrm{V/K}$ at 300 K, and its absolute value is maximal among the investigated samples. The Seebeck coefficients of $Co_{1-y}Ni_yGe$ are n-type and the absolute value of S monotonously decreases with increasing y. S shifts toward a positive side with Fe doping and

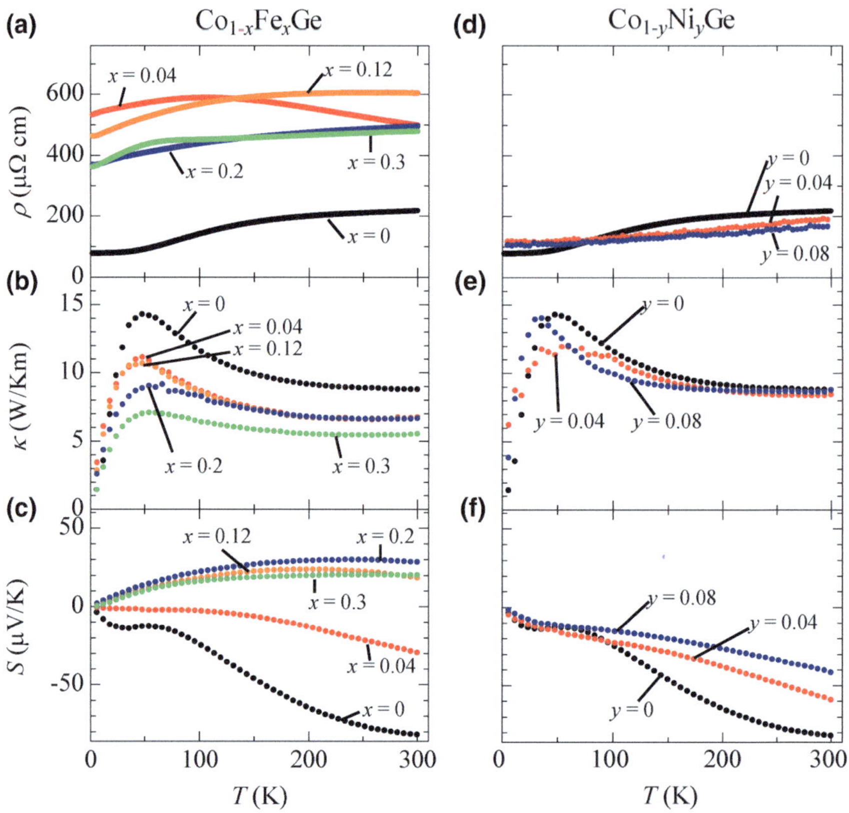

Fig. 6.5 Temperature dependence of resistivity ρ, thermal conductivity κ and Seebeck coefficient S in $Co_{1-x}Fe_xGe$ [panels (**a**)–(**c**)] and $Co_{1-y}Ni_yGe$ [panels (**d**)–(**f**)], respectively. Reprinted with permission from Ref. [20]. Copyright 2012, American Institute of Physics

above $x = 0.12$ shows a positive value below 300 K. The largest positive S (28 μV/K at 300 K) is achieved at $x = 0.2$, and then S decreases with further increasing x (or decreasing band filling).

On the basis of the rigid band approximation (RBA), the Fermi level (E_F) of $Co_{1-x}Fe_xGe$ (hole doping) and $Co_{1-y}Ni_yGe$ (electron doping) are shown in Fig. 6.2c. The intuitive explanation described above in terms of the asymmetric band structure around Γ point (the inset of Fig. 6.2c) predicts both in the magnitude and the abrupt p-n transition of S with changing the electron filling. This speculation is indeed observed as shown in Fig. 6.6b, which describes the comparison between the calculated and experimental results of S at 300 K. The calculated results semi-quantitatively reproduce the band filling dependence of the observed thermoelectric effect, both in the magnitude and the abrupt p-n transition of S. While the negatively-maximal S in the experimental result is realized at CoGe, that in the calculation is at $y = 0.04$. This implies some discrepancy between the calculated and actual band

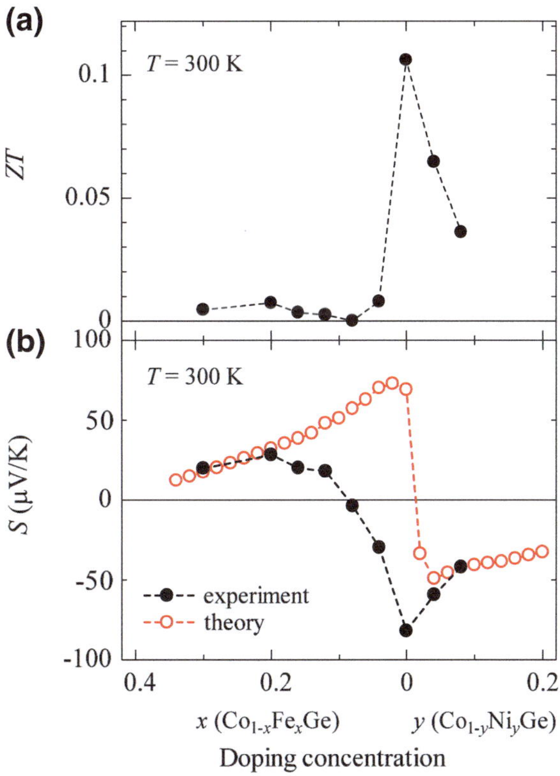

Fig. 6.6 Band filling dependence of **a** figure of merit ZT and **b** Seebeck coefficient S at 300 K. The *solid* and *open circles* indicate the experimental and theoretical values of S, respectively. The *dashed lines* are the guide to the eyes. Reprinted with permission from Ref. [20]. Copyright 2012, American Institute of Physics

structures/parameters, in particular around E_F. As DOS around E_F of CoGe is small forming the pseudogap, a slight deviation of calculated band structure from the reality should cause the shift of E_F from the level for the minimum S. The observed S tends to be negatively shifted from the theoretical prediction; the negative maximal peak of S in the calculation is enhanced while the positive one is totally suppressed in reality. That may indicate the need for taking an appropriate energy dependence of relaxation time in calculation.

Finally, the figure of merit ZT for all the samples is shown in Fig. 6.6a. CoGe shows the maximum figure of merit $ZT \approx 0.11$, as large as that of CoSi, owing to its metallic low resistivity and large negative S. The Fermi level of CoGe ($x = y = 0$) is already close to the optimal as seen from the doping systematics (Fig. 6.6a). This result indicates that a well-designed management to lower the thermal conductivity, such as the solid solution or nano-scale mixture with CoSi and the superlattice formation, may realize the further improvement of the thermoelectric efficiency.

6.3 Summary

We have measured the band-filling dependence of thermoelectric properties in B20-type CoGe, which can be synthesized by a high-pressure synthesis method. The good accordance of the observed Seebeck coefficient with the calculated one indicates that the asymmetric band structure around E_F composed of the Dirac cone and the flat band with sharp bends is the origin for the large magnitude of positive or negative S. CoGe shows the largest figure of merit $ZT \approx 0.11$ at 300 K and upward tendency with increasing temperature. High thermoelectric efficiency can be achieved in the metallic systems with Dirac electron states due to their "unusual" energy dependence of DOS.

References

1. G.D. Mahan, Solid State Phys. **51**, 81 (1998)
2. Y. Nishino, S. Deguchi, U. Mizutani, Phys. Rev. B **74**, 115115 (2006)
3. H.J. van Daal, P.B. van Aken, K.H.J. Buschow, Phys. Lett. A **49**, 246 (1974)
4. R.J. Gambino, W.D. Grobman, A.M. Toxen, Appl. Phys. Lett. **22**, 506 (1973)
5. I. Terasaki, Y. Sasago, K. Uchinokura, Phys. Rev. B **56**, R12685 (1997)
6. N.W. Aschcroft, N.D. Mermin, *Solid State Physics* (Thomson Learning, London, 1976)
7. K. Kuroki, R. Arita, J. Phys. Soc. Jpn. **76**, 083707 (2007)
8. G.Y. Guo, G.A. Botton, Y. Nishino, J. Phys. Condens. Matter **10**, L119 (1998)
9. W. Koshibae, K. Tsutsui, S. Maekawa, Phys. Rev. B **62**, 6869 (2000)
10. S. Asanabe, D. Shinoda, Y. Sasaki, Phys. Rev. **134**, A774 (1964)
11. D.J. McNeill, R.M. Ware, Br. J. Appl. Phys. **15**, 1517 (1964)
12. A. Sakai, F. Ishii, Y. Onose, Y. Tomioka, S. Yotsuhashi, H. Adachi, N. Nagaosa, Y. Tokura, J. Phys. Soc. Jpn. **76**, 093601 (2007)
13. C.S. Lue, Y.K. Kuo, C.L. Huang, W.J. Lai, Phys. Rev. B **69**, 125111 (2004)
14. C.C. Li, W.L. Ren, L.T. Zhang, K. Itoh, J.S. Wu, J. Appl. Phys. **98**, 063706 (2005)
15. W.L. Ren, C.C. Li, L.T. Zhang, K. Ito, W.J. Lai, J. Alloys Compd. **392**, 50 (2005)
16. Y.K. Kuo, K.M. Sivakumar, S.J. Huang, C.S. Lue, J. Appl. Phys. **98**, 123510 (2005)
17. E. Skoug, C. Zhou, Y. Pei, D.T. Morelli, Appl. Phys. Lett. **94**, 022115 (2009)
18. F. Wald, S.J. Michalik, J. Less-Common Met. **24**, 277 (1971)
19. H. Takizawa, T. Sato, T. Endo, M. Shimada, J. Solid State Chem. **73**, 40 (1988)
20. N. Kanazawa, Y. Onose, Y. Shiomi, S. Ishiwata, Y. Tokura, Appl. Phys. Lett. **100**, 093902 (2012)
21. P. Blaha, K. Schwarz, G. Madsen, D. Kvasnicka, J. Luitz, *An Augmented Plane Wave + Local Orbitals Program for Calculating Crystal Properties* (Karlheinz Schwarz, Techn. Universität Wien, Austria, 2001)
22. J.P. Perdew, K. Burke, M. Ernzerhof, Phys. Rev. Lett. **77**, 3865 (1996)
23. G.K.H. Madsen, D.J. Singh, Comput. Phys. Commun. **175**, 67 (2006)

Chapter 7
Conclusion

Motivated by the various interesting phenomena observed in B20-type silicon compounds, we have studied B20-type transition-metal germanium compounds mainly focusing on skyrmion magnetic structure and Dirac band structure. Although most of the B20-type germanides cannot be synthesized at ambient pressure, we have succeeded in growing a series of B20-type compounds of $Mn_{1-x}Fe_xGe$ and $Fe_{1-y}Co_yGe$ by utilizing various crystal growth methods, such as high-pressure synthesis and molecular beam epitaxy.

Through screening of their physical properties, we have found that the system can be categorized into three different phases: (i) short-period helimagnets with strong spin-orbit interaction, (ii) long-period helimagnets, and (iii) non-magnetic phase. The first category is a newly identified phase, which dose not appear in B20-type silicides. Three mother compounds, i.e. MnGe, FeGe, and CoGe, are closely investigated as typical examples of the three phases.

In Chap. 4, we have studied MnGe by Hall resistivity measurements and neutron scattering experiments to determine its magnetic structure. A novel spin texture of 3D skyrmion crystal state, which is a superposition of three orthogonal helical structure, is possibly realized in a wide temperature-magnetic-field region. This topologically winding spin texture produces characteristic emergent magnetic field (gauge flux due to Berry phase), appearing as an unusual Hall effect (topological Hall effect). Strong spin-orbit coupling may play an important role in the emergence of the 3D spin crystal phase via antisymmetric spin exchange interactions like Dzyaloshinsky-Moriya interaction. The distribution of emergent magnetic field, which conduction electrons feel, corresponds to that induced by pairs of magnetic monopole and anti-monopole. Searching further novel electromagnetic responses in MnGe are future challenges.

In Chap. 5, we have studied nanoscale patterned electronic circuits with various sizes (32 nm–10 μm) in B20-type FeGe thin film. Among the B20-magnets, FeGe shows the highest transition temperature of 280 K, which is close to possible practical application. It is of capital importance to understand how skyrmions behaves in the device made of the model compound. Skyrmion formation was successfully detected

N. Kanazawa, *Charge and Heat Transport Phenomena in Electronic and Spin Structures in B20-type Compounds*, Springer Theses,
DOI 10.1007/978-4-431-55660-2_7

as topological Hall effect in the Hall signal. Skyrmion structure remains even in the one-skyrmion-size device. We can say that prototypical circuit for reading out skyrmion existence is realized.

In Chap. 6, we have measured thermoelectric performance of CoGe with Dirac electrons. As the consequence of unusual DOS spectrum $\propto E^2$, a large thermopower can be realized simultaneously with a large electrical conductivity, leading to good figure of merit $ZT \approx 1.1$.

Just substituting the sites of group-14 elements, various dramatic changes occur in the magnetic/electronic structures, producing novel electromagnetic responses as seen above. Some compositions which cannot crystalize into B20-type at ambient pressure and remain unexploited might be a treasure box of new physics.

Curriculum Vitae

Naoya Kanazawa

Affiliation:

Department of Applied Physics, School of Engineering, the University of Tokyo.

Postal Address:

Engineering Bldg. 6, Room #125, the University of Tokyo, Hongo 7-3-1, Bunkyo-ku, Tokyo 113-8656, JAPAN
E-mail: kanazawa@ap.t.u-tokyo.ac.jp

Appointments:

- Research Associate, Tokura Laboratory, the University of Tokyo (April 2014–)
- JSPS Research Fellow (DC1), (April 2011–March 2014)

Education:

- Ph. D. in Applied Physics, the University of Tokyo (2011–2014), advisor: Professor Yoshinori Tokura
- M. S. in Applied Physics, the University of Tokyo (2009–2011), advisor: Professor Yoshinori Tokura
- B. S. in Applied Physics, the University of Tokyo (2005–2009), advisor: Professor Yoshinori Tokura

Current Research Interests:

Topological spin textures and their quantum transport phenomena

Publication list:

1. X.Z. Yu, Y. Onose, N. Kanazawa, J.H. Park, J.H. Han, Y. Matsui, N. Nagaosa, Y. Tokura, Real-space observation of a two-dimensional skyrmion crystal. Nature **465**, 901 (2010).

N. Kanazawa, *Charge and Heat Transport Phenomena in Electronic and Spin Structures in B20-type Compounds*, Springer Theses,
DOI 10.1007/978-4-431-55660-2

2. X.Z. Yu, N. Kanazawa, Y. Onose, K. Kimoto, W.Z. Zhang, S. Ishiwata, Y. Matsui, Y. Tokura, Near room-temperature formation of a skyrmion crystal in thin-films of the helimagnet FeGe. Nat. Mater. **10**, 106 (2011).

3. N. Kanazawa, Y. Onose, T. Arima, D. Okuyama, K. Ohoyama, S. Wakimoto, K. Kakurai, S. Ishiwata, Y. Tokura, Large topological hall effect in a short-period helimagnet MnGe. Phys. Rev. Lett. **106**, 156603 (2011).

4. A. Tonomura, X. Yu, K. Yanagisawa, T. Matsuda, Y. Onose, N. Kanazawa, H.S. Park, Y. Tokura, Real-space observation of skyrmion lattice in helimagnet MnSi thin samples. Nano Lett. **12**, 1673 (2012).

5. N. Kanazawa, Y. Onose, Y. Shiomi, S. Ishiwata, Y. Tokura, Band-filling dependence of thermoelectric properties in B20-type CoGe. Appl. Phys. Lett. **100**, 093902 (2012).

6. X.Z. Yu, N. Kanazawa, W.Z. Zhang, T. Nagai, T. Hara, K. Kimoto, Y. Matsui, Y. Onose, Y. Tokura, Skyrmion flow near room temperature in an ultralow current density. Nat. Commun. **3**, 988 (2012).

7. N. Kanazawa, J.-H. Kim, D.S. Inosov, J.S. White, N. Egetenmeyer, J.L. Gavilano, S. Ishiwata, Y. Onose, T. Arima, B. Keimer, Y. Tokura, Possible skyrmion-lattice ground state in the B20 chiral-lattice magnet MnGe as seen via small-angle neutron scattering. Phys. Rev. B **86**, 134425 (2012).

8. Y. Li, N. Kanazawa, X.Z. Yu, A. Tsukazaki, M. Kawasaki, M. Ichikawa, X.F. Jin, F. Kagawa, Y. Tokura, Robust formation of skyrmions and topological hall effect anomaly in epitaxial thin films of MnSi. Phys. Rev. Lett. **110**, 117202 (2013).

9. M. Kubota, Y. Tokunaga, N. Kanazawa, F. Kagawa, Y. Tokura, M. Kawasaki, Miniature hall sensor integrated on a magnetic thin film for detecting domain wall motion. J. Appl. Phys. **114**, 053909 (2013).

10. Y. Shiomi, N. Kanazawa, K. Shibata, Y. Onose, Y. Tokura, Topological nernst effect in a three-dimensional skyrmion-lattice phase. Phys. Rev. B **88**, 064409 (2013).

11. D. Morikawa, K. Shibata, N. Kanazawa, X.Z. Yu, Y. Tokura, Crystal chirality and skyrmion helicity in MnSi and (Fe, Co)Si as determined by transmission electron microscopy. Phys. Rev. B **88**, 024408 (2013).

12. K. Shibata, X.Z. Yu, T. Hara, D. Morikawa, N. Kanazawa, K. Kimoto, S. Ishiwata, Y. Matsui, Y. Tokura, Towards control of skyrmion crystals in helimagnetic alloys by spin-orbit coupling. Nat. Nanotechnol. **8**, 723 (2013).

13. M. Mochizuki, X.Z. Yu, S. Seki, N. Kanazawa, W. Koshibae, J. Zang, M. Mostovoy, Y. Tokura, N. Nagaosa, Thermally driven ratchet motion of skyrmion microcrystal and topological magnon hall effect. Nat. Mater. **13**, 241 (2014).

14. T. Yokouchi, N. Kanazawa, A. Tsukazaki, Y. Kozuka, M. Kawasaki, M. Ichikawa, F. Kagawa, Y. Tokura, Skyrmion stability in thin films of $Mn_{1-x}Fe_xSi$ as investigated by topological hall effect. Phys. Rev. B **89**, 064416 (2014).

15. H.S. Park, X. Yu, S. Aizawa, T. Tanigaki, T. Akashi, Y. Takahashi, T. Matsuda, N. Kanazawa, Y. Onose, D. Shindo, A. Tonomura, Y. Tokura, Observation of the magnetic flux and three-dimensional structure of skyrmion lattices by electron holography. Nat. Nanotechnol. **9**, 337 (2014).

16. N. Kanazawa, M. Kubota, A. Tsukazaki, Y. Kozuka, K.S. Takahashi, M. Kawasaki, M. Ichikawa, F. Kagawa, Y. Tokura, Discretized topological hall effect emerging from skyrmions in constricted geometry. Phys. Rev. B **91**, 041122(R) (2015).

MIX
Papier aus verantwortungsvollen Quellen
Paper from responsible sources
FSC® C105338

If you have any concerns about our products,
you can contact us on
ProductSafety@springernature.com

In case Publisher is established outside the EU,
the EU authorized representative is:
Springer Nature Customer Service Center GmbH
Europaplatz 3, 69115 Heidelberg, Germany

Printed by Libri Plureos GmbH
in Hamburg, Germany